FUNDED!

FUNDED!

SUCCESSFUL GRANTWRITING FOR YOUR NONPROFIT

Richard Hoefer

OXFORD
UNIVERSITY PRESS

Oxford University Press is a department of the University of Oxford. It furthers the University's objective of excellence in research, scholarship, and education by publishing worldwide. Oxford is a registered trade mark of Oxford University Press in the UK and certain other countries.

Published in the United States of America by Oxford University Press
198 Madison Avenue, New York, NY 10016, United States of America.

Library of Congress Cataloging-in-Publication Data
Names: Hoefer, Richard, author.
Title: Funded! : successful grantwriting for your nonprofit / Richard Hoefer.
Description: New York, NY : Oxford University Press, 2017. |
Includes bibliographical references and index.
Identifiers: LCCN 2016055689 (print) | LCCN 2017013336 (ebook) |
ISBN 9780190687267 (updf) | ISBN 9780190681876 (alk. paper)
Subjects: LCSH: Proposal writing for grants. | Nonprofit organizations—Finance.
Classification: LCC HG177 (ebook) | LCC HG177 .H635 2017 (print) |
DDC 658.15/224—dc23
LC record available at https://lccn.loc.gov/2016055689

*To Paula Norine Homer and Sharon Elizabeth Hoefer,
bringers of music and joy to my life.*

CONTENTS

PREFACE

Congratulations on being interested in becoming a grant writer. It is a very rewarding job but also one that has no clear career entry point. It's no wonder that most people have little idea how to get started as a grant writer. This book introduces you to some of the current realities of the field and what you need to know to join us in our career.

Just so you know where I'm coming from and why this book means so much to me, let me tell you about my introduction to being a grantwriter.

In the fall of 1980, I was a graduate student beginning an internship where I would learn to write grants (so you can see I've been at this a long time!).

On Monday, my first day, I met my new supervisor, Judy. She handed me two pencils and a yellow legal pad. Then she gave me a stuffed 10" × 13" brown manila envelope.

Judy instructed me, "I'd like you to start reading these materials. You'll see that there's a request for proposals from the Labor Department in there, and a couple of pages of notes that you can use to develop the 20-page, single-spaced, application."

I smiled and told Judy, "Wow! This is exactly what I'm looking for in my internship." Her reply: "Great! Just be sure it's done by Thursday at noon, because I want to look it over before we have to overnight it to DC by 5 p.m."

My joy turned to panic in a heartbeat. I must have looked as frightened as I felt.

Remember, this is on Monday, and I have classes and other things to do before Thursday noon.

Plus, I've never written a grant before. I gasped, "That's not much time, is it?"

Judy just flatly responded, "No, it's not. You better get started."

I thought she would continue with, "but we will all pitch in and help you," but she didn't. She just ended her sentence with a cold, hard period.

But then she said, "When you are done with that, here is this other folder. This is another grant that we're working on." It was a big envelope, like the first, also stuffed with forms and paper. Then she smiled and continued. "But don't worry. It's not due until the next Friday, so you'll have over a week to finish it."

I was having second thoughts about this internship.

But then the conversation continued. (It wasn't really a conversation, because I wasn't saying much.) Judy consoled me by saying, "You seem a bit upset. But you know what? Don't worry too much about it, Rick. I don't expect that we're going to get either one of these grants. But I figure it's worth a shot. Just do your best."

To get a second opinion, I went to my faculty advisor and asked him, "Professor Zimmerman, isn't this is an unusual way to teach people how to do grantwriting?"

He told me, "No, it's not unusual at all. That's just the way it's done. You just do it. You learn by doing it."

(By the way, Judy was right. We didn't get either one of those grants.)

That was my introduction to writing grant proposals. No training, no class, very little help—just a trial by fire. I knew in my heart that this was NOT the right way to train grantwriters!

Still, something good did come out of that initial grantwriting experience. Because of what I went through that day, I dedicated myself to learning everything I could about writing grant proposals.

I knew that if I was going to become a decent grantwriter, I was going to have to work out the details on my own. I was going to have to read everything I could and I was going to have to work through lunches, dinners, and late into the night sometimes.

Did I make mistakes along the way? I certainly did. (If we ever have the chance to talk, ask me about my encounter with the associate dean of the School of Business. I still shudder when I think about him.)

But I also learned a lot about what I needed to know.

By the end of my 8-month internship, I had written successful grants for hundreds of thousands of dollars, and I was hired as director of fundraising for that organization until I left the area.

And since then, I've gotten better. On my own, or in teams, I can count many millions of dollars' worth of funding over the course of my grantwriting career. I've also taught other people how to write grants, in person and through the University of Texas at Arlington's continuing education program.

Before we go any further, let's figure out how to spell the term. Is it "grantwriting," "grant writing," or "grant-writing"? My preference is to make it all one word with no punctuation. You'll see it different ways with different authors, but in this book, it's going to be spelled "grantwriting."

I hope you'll be like others who have been able to learn from my experiences. This book in your hands now is an introduction to the profession of grantwriting, and to the art and science of grantwriting.

Use this knowledge to help the causes, organizations, and client populations that are dear to you. You can make a huge difference to so many people—take it one step at a time and get better over time. The world needs you and your skills.

Sincerely,
Dr. Richard Hoefer

P.S. Subscribe to my newsletter list to stay up on the latest trends, ideas, and practices. Go to www.richardhoefer.com, enter your e-mail address, and you're set!

FUNDED!

CHAPTER 1

GRANTWRITING AND YOU

I was once at a going away party for several nonprofit employees. These were good workers, had been at the agency many years, and were not leaving because of other job offers. I sat down next to a person I didn't recognize who seemed to be on the verge of tears. I asked her if I could help, somehow. She looked at me, with tears welling in her eyes, shaking her head ever so slowly. "If only I were better at my job this wouldn't be happening. These people, my friends, are being laid off because the grant I wrote wasn't funded. We have no money to continue their program. I just want to be a better grantwriter, but I don't know how!" How well would *you* handle this situation, even knowing that not even the best grantwriters have a 100 percent success rate?

Being a grantwriter is an exciting job: the work is constantly changing, you meet with a variety of people, and using your skills well can literally be the difference between your employer thriving or dying. This is not a job for the easily discouraged, however, or for those seeking a 9–5 job. It IS a job for those who wish to make a difference in the lives of clients on a vast scale. When you can write successful proposals you have a skill that is vital to nonprofit organizations.

Of course as a grantwriter you do more than just write. In the process of developing a proposal you have to learn how to gather information from many sources, keep current on cutting-edge ideas relating to various topics you may write about, learn the tricks of the trade to have top-notch proposals, use excellent people skills to ensure you have harmonious working relationships with the people in the agency you're writing for, and many other tasks.

Being a grantwriter can bring you great personal satisfaction as you help an agency achieve its goals by helping ensure it is adequately funded. Many grantwriters simply enjoy the intellectual challenge that putting together a proposal represents. It's also a wonderful feeling to be awarded a grant, as a validation of your work and skills.

This chapter covers a lot of ground: the majority of it is an assessment to help you understand how you and grantwriting might fit together. We then look at the skills, education, and training you should have as a grantwriter, at the pre-writing, writing, and post-writing stages of proposal development. Once you know what skills are important, you'll want to explore how to get those skills and how to get your first experience as a grantwriter. Many people want to know more about being a grantwriting employee compared to being a freelancer, so that topic is explored as well.

ASSESSING WHETHER GRANTWRITING IS A GOOD PROFESSION FOR YOU

All of us have special skills and abilities that help us be good at some jobs and not at others. In this vein, it is important to understand that being a grant writer is not a job for everyone! It takes a certain disposition and set of attributes in order to make it a good fit for you. This is *not* meant to be a discouraging comment but simply to say if you don't fit a particular job *right now,* it's good to know the challenges you may face to become excellent at the tasks required.

Some people decide they want to become a grantwriter because they have heard it can pay well (see myth 4, in chapter 2). Other people decide to write grants because they like to write. This is important, of course, but not sufficient. Some people get into proposal writing because they've been told to do it by their supervisor or in order to get a job in the nonprofit sector.

None of these reasons are "bad"—but they may not be enough to keep a person in the position for the long haul. One of the facts that people who are just getting started don't know is that there is a large turnover in most grantwriting and development positions—the average tenure in a grantwriting position is just about two years. The next few pages present a self-assessment of where you currently stand in terms of having grantwriting be a good fit for you. Take the time to complete it in a thoughtful and sincere way—it will help you decide if this is the right profession for you or not.

It may be helpful to know that grantwriters often specialize in one area of writing. If you're going to move to the top of the field, you'll want to know a lot about a small area of grantwriting. That way you can become very well known and also get to know the main foundations and government agencies that provide funding in that area. For instance, if you were to specialize in the area of human services, that would still be broad in scope. At least you wouldn't try to write grants to support an opera house, but human services are still a very broad field of endeavor. True specialization and the ability to become an expert come from narrowing your focus at least somewhat. If you specialized in writing grants to provide services for HIV/AIDS treatment and services, as well as prevention, you would have a viable career. As funders change their priorities within the field, you could still broaden your focus but keep a firm foundation on what you started with. In this way it can be very advantageous to get your degree in an area where you can write grants. You may find a bachelor's degree in sociology with lots of writing required is a very good preparation for becoming a grantwriter, especially if you combine that preparation with an advanced degree in social work.

WHAT THIS ASSESSMENT IS DESIGNED TO DO

This assessment is designed to help you determine if you CURRENTLY have what it takes to become a skilled and successful grant writer. You'll be asked questions to help *you* decide if this is for you. While many different types of people ARE grant writers, not all of them thrive in the position. Many happy and successful grant writers share similar traits and personalities. But even people who have the correct temperament and background may not have all the skills needed yet. So this assessment looks also at the skills you currently have and, based on your answers, will suggest areas where you may wish to augment your skills and knowledge.

This assessment is based on practice wisdom rather than on scientific verification techniques—so if you don't get the results you think you should have gotten, and you still want to work on becoming a grant writer, don't despair! Going through the questions will raise issues that you may find useful to think about. You'll learn a lot about grantwriting, too.

BEING A GRANTWRITER ASSESSMENT INSTRUMENT

The assessment is divided into several parts. Do all the parts, in the order they are written, answering all the questions, in one sitting. If this is not possible, do as much as you can, and return to complete the assessment as soon as you can. This will give you the "truest" reflection of your current situation.

ASSESSMENT PART 1: TEMPERAMENT

To what extent are these statements true for YOU? Check the most appropriate response in each row.

	Very Untrue	**Untrue**	**Neither True Nor Untrue**	**True**	**Very True**
I am motivated by idealism					
I am very pragmatic and practical					
I think it is fun to track down specific facts and figures to support my ideas					
I enjoy reading dense technical reports to learn the latest in my field					
I can read and easily remember mundane details in a lengthy document					
I make it a priority to read a variety of nonfiction materials, such as newspapers, weekly and monthly magazines, websites, and official documents					
It doesn't bother me to work in a stressful environment					
I encourage honest feedback even when it is negative, if it will lead me to being able to do a better job					
Even big setbacks don't get me down for long					
If an idea doesn't work out at first, I try to use the best parts of it in a different way					
Number of boxes checked in each column (0 to 10 possible)					
Multiply number boxes checked by:	0	1	2	3	4
Column Subtotal					
PART 1 Score Sum all five columns for total temperament score: (0 to 40 possible)					

ASSESSMENT PART 2: WRITING

To what extent are these statements true for YOU? Check the most appropriate response in each row.

	Very Untrue	Untrue	Neither True Nor Untrue	True	Very True
Proposals should showcase my writing style					
Grants should hardly ever use bulleted lists					
Grant proposals should test theories more than achieve concrete service goals					
It isn't easy for me to use dry facts and figures to engage readers' imaginations					
I firmly believe that getting things 80–90% right in a grant application allows me to get more done; anything more is overkill					
Jargon and technical words are absolutely vital to show I understand grant objectives					
I believe that technical documents such as grant proposals are inherently dull					
"Marketing" and "selling" are not appropriate approaches to grantwriting					
It is more important to show how organizational or personal goals will be achieved than funder goals					
I carefully construct my sentences and paragraphs even if it takes more time than I have					
Number of boxes checked in each column (0 to 10 possible)					
Multiply number boxes checked by:	4	3	2	1	0
Column Subtotal					
PART 2 Score Sum all five columns for total writing score: (0 to 40 possible)					

ASSESSMENT PART 3: OTHER GRANT WRITING ELEMENTS

To what extent are these statements true for YOU? Check the most appropriate response in each row.

	Very Untrue	Untrue	Neither True Nor Untrue	True	Very True
I have a strong professional identity					
I like developing lists of things to do and checking items off					
I enjoy interacting with subject area experts to stay up on the latest developments					
I believe program staff members should be consulted to better understand their viewpoints about possible grant proposals					
My organization skills are exceptionally strong					
Number of boxes checked in each column (0 to 5 possible)					
Multiply number boxes checked by:	0	1	2	3	4
Column Subtotal					
PART 3 Score Sum all five columns for total other grantwriting elements score: (0 to 20 possible)					

Part 1 Score	
Part 2 Score	
Part 3 Score	
TOTAL Score	

HOW TO INTERPRET YOUR SCORE

As mentioned earlier, this assessment is designed to help you determine for yourself whether you have what it takes to become a skilled and successful grant writer. You might have noticed that the questions weren't totally about your knowledge of how to do things required to write a grant (such as develop a logic model or put together a budget). Although these are things that you absolutely must know, they are relatively easily taught and learned. The focus here is on items such as what you enjoy doing, what you are good at, how you process and retain information, what your attitude is toward your work, and the like.

With that in mind, look again at your scores. The higher your score, the more likely it is that you will do well as a grant writer because you have the right personality traits and approach to the job. My experience indicates that a minimum score of 70 is a good starting point.

But suppose that you don't have a score of 70 and you still think that you would be a good grant writer. If this is true, it is important to look at the subtotals. Is one of the three areas very low? If so, you have pinpointed an area where you may need some additional preparation.

For example, if you have a low score in "Part 1: Temperament" it may be because you don't currently enjoy reading technical reports to stay up on the latest trends and ideas in your field, and you don't make it a priority to read a variety of nonfiction materials. You also may not have thought that you remember details in a lengthy document. Scoring low on just these three items can bring down your score quite a bit. But with knowledge and practice, these are things that you can change about how you live and work.

If you have a low score in "Part 2: Writing" it may be that you currently have a more typical academic style of writing where you have to show command of the esoteric jargon of your field, you would almost never use bulleted lists, and you would want to test theory rather than provide and evaluate services. All of these aspects of your writing can be changed so you can become a much more vibrant and practically oriented grant writer.

Another issue that arises around writing is that some potential grant writers have what I call "English major's malady"—a sense that their personal style is an overarching concern and that getting the writing "exactly so" is more important than getting the grant turned in on time. Again, these attitudes can be changed to better fit the demands of the grantwriting job.

The last set of questions, "Part 3: Other Grantwriting Elements," is a more varied group of items. All of them relate to ways you like to act. It may be that you have not been really sure of what kinds of things a good grant writer should like to do or what skills they should be able to learn. Once these have been pointed out to you, you now have a great deal more information than you did before so that you can make a more informed decision as to whether you should pursue a career in grantwriting.

WHAT SHOULD YOU DO NEXT?

If you believe you have what it takes to become a skilled grant writer, and you are excited to build on your current attitudes, beliefs, skills and knowledge, then you should move forward with your dream. The rest of this book can help you launch your career with solid knowledge and an action plan. Someone who can consistently present new ideas to potential funders in an engaging and convincing way can truly help make the world a better place. You'll also be in demand for your skills and ability to raise significant amounts of funds for an organization.

WHAT SKILLS/EDUCATION/TRAINING SHOULD YOU HAVE?

As reflected in the assessment, grantwriters need a variety of skills. There is no "best way" to acquire these, but it is important to know what you're in search of before you start the process, or to know where you could use a bit more background.

Let's first look at the skills that are needed as you move through the typical grantwriting process. Then we'll talk about the desirability of having some specific substantive knowledge and specializing in a particular field of nonprofits.

PRE-WRITING STAGE

Even before you begin to write a grant, during the "pre-writing stage", you need to prepare yourself by knowing your organization and finding grantwriting opportunities.

KNOW YOUR ORGANIZATION

While it may not be obvious, you cannot write a proposal that is going to be competitive if you don't know the organization that you're working for. In some ways, your proposal is like a personal recommendation. You are telling the funder that the organization you're working for is a good bet for spending the money and getting results. Just as with any recommendation, you have to really know the strengths and weaknesses of who you're writing a referral for to be convincing.

Here are some things to know about your organization:

- *History*—How did it start and why?
- *Current projects*—What is the organization currently known for? What type of programs does the organization currently run?
- *Personnel*—Who is on the leadership team, particularly the executive director? Who is on the board of directors? What types of staff are on board already? What are their skills?
- *Systems*—How strong are the financial systems in the organization? Can the agency handle the reporting and evaluation requirements of the grant?
- *Aspirations*—Where would the organization like to be in 5 or 10 years? What does it want to be known for?

FIND OPPORTUNITIES

Before you begin to write a proposal, you must have a proposal to write for! Thus, the skill you'll use earliest in the process is searching for and finding appropriate funding opportunities, both in the world of foundations and among government funding opportunities. The interesting thing about grantwriting is that almost all organizations giving away money tell you what they want to fund. Of course, you need to be savvy about how to decipher the information they provide, but it is their interest as well as yours to make a close match before you invest the time to write a proposal and they take the time to review it.

WRITING STAGE

Once you've completed the pre-writing stage, you move to actually preparing your grant proposal. The skills you need here include research, program design, information gathering and collaboration, and written communication.

RESEARCH

Research is involved in almost all parts of grantwriting. One of the key elements of writing a proposal is making a case that a need or problem exists. This takes considerable effort and a firm set of facts to be convincing. Research is also vital for offering a solution to the problem. What will help make that problem decrease or disappear? What scientific research do you base that on? Designing and implementing a program can be challenging. What information do you have that shows your proposed program will work with the population you have in mind? How strong is the evidence in favor of this approach compared with another approach that also looks good? As the grantwriter, you'll be expected to have ideas that are solid and link back to credible evidence.

PROGRAM DESIGN

One of the beautiful aspects of being a grantwriter is the opportunity to create a program. One of the most horrific aspects of being a grantwriter is the need to put together a program. A program is a set of activities to bring together funds, people, and other elements, such as information and equipment, to solve problems. For example, if you want to solve the problem of 6- to 8-year-old children not knowing how to read, you might design a program that involves volunteers reading books to children and helping them learn to read for themselves in an after-school program. As a grantwriter, you need to imagine how the set of activities should be put together in order for it to be effective in solving the problem. As you can guess, you have to do your homework to find out what has been tried in the past, with what level of success, and how previous efforts can be improved and adapted to fit your intended client population.

INFORMATION GATHERING/COLLABORATION

A grantwriter almost never works on a grant entirely alone. So a skill that you need to have is the ability to gather ideas from other people to determine what is feasible and what is desirable. For instance, as a grantwriter you may come across some program ideas that are research-based, have worked with your organization's desired client population, and so on, but you may find that the director of programming isn't enthusiastic about the idea. This may be for many different reasons, but if you don't have support from the key staff members of the agency, you shouldn't go ahead with the proposal.

You also need the ability to take different people's ideas and work them into a single proposal. You are the one who understands what is required in the proposal better than anyone else so you need to guide the discussions and provide the impetus for collaboration among all the people who are involved, from the front-line staff to the CEO. As you listen to everyone, you also have to choose what to accept and what to reject without making people mad or disengaged from the process.

WRITTEN COMMUNICATION

If you want to be a good grantwriter you absolutely must be able to write well. You don't need a master's degree in English, but you have to be able to convey ideas through writing extremely well. Your command of formal or academic English should be strong, simply because that is the language that funders use and expect from their grantees.

As noted earlier, grantwriting is a form of technical writing, meaning that it is not about "self-expression" or being "literary." Grantwriting's highest expression is to paint a picture of an ongoing need and then present a viable solution to that need. It should be presented in a way that is both scientific and humanistic. Funders want both to see and feel the problem while knowing that indisputable information backs up the proposal.

POST-WRITING STAGE

After the proposal draft is completed and approved by the board of directors and the CEO of the organization, it needs to be transmitted to the foundation or government agency that is targeted to receive it. Sometimes this will be an electronic upload through the federal government's electronic submission system, which can be a daunting task all by itself. Other times this may be a mailed copy of the proposal going straight to a foundation.

Federal grants have a strict deadline. If the request for proposals indicates that no proposals will be accepted after 5 p.m., you can be sure that all proposals that are uploaded after that time will be rejected. If this happens to you, you will feel terrible. All your work and the efforts of those who assisted you will be, for the time being, wasted. While you may be able to recycle some of the proposal into another opportunity, there is no guarantee of that.

If your proposal scores well, you will likely be invited to communicate with the funder. In this case you will need to have strong oral communication skills to match your written skills. Even if you are interviewed or have a site visit, the grant may not be awarded to your organization. Many things can trip up the award, including a sudden decrease in the amount of funding the grantor has to hand out, or an unfavorable review of the organization's internal systems. It is up to the grantwriter and others on the administrative team to organize a coherent presentation and be truly competent in running an agency.

WHERE DO I GET GRANTWRITING SKILLS?

There is a legitimate question to ask: Where can I get these skills? Do I need a particular degree or certification? Let's look at these issues.

Many of the skills needed for being a successful grantwriter can be learned in college. Research and writing skills top the list of college-related abilities. Working on projects in groups is another skill often experienced and honed in college. The other skills listed above are usefully obtained through a college education, even if there is no course on grantwriting by itself.

Probably the skill least often found in coursework is how to find government and foundation grant opportunities. This can be learned from trainings for grantwriters, but even then it is not always taught well. The Foundation Center has course offerings (for a fee) that can help you with researching foundations. They teach you by using their product, Foundation Directory Online, which is an excellent source of information on foundations. The directory is pretty expensive, but is often available for free in public libraries. Having a good mentor is useful, but librarians can often provide a great deal of help.

Learning how to design a program to put in your grant proposal is also a fairly rare topic outside of a few professional degrees, such as social work. It is vital, however, as every grant proposal has to have a solution to the problem it discusses.

These skills are more commonly taught at the master's level than at the bachelor's level. Many different academic fields have courses on grantwriting, such as social work, nonprofit management, and public administration. But if you are interested in grantwriting, you can obtain the necessary skills from anywhere and you can try out the life of a grantwriter. But how do you get your foot in the door to write that first proposal?

HOW DO I GET MY EXPERIENCE AS A GRANTWRITER?

The world of nonprofit grantwriting is similar in some ways to the arena of acting. You've got to prove yourself in some way before you're going to be offered a job. In acting, these are called "auditions," where you perform for a short time to see if you match the director's vision for a role in the next production. You can't just walk in off the street as a new actor and say you're ready for the part! To get a job as a grantwriter there are a number of things that will help that are not directly related to the grantwriting job itself.

First, it will be beneficial if you've had experience in the nonprofit world. While each organization has its own culture, the nonprofit sector as a whole has some strong ideas about how things should be done. If you haven't been around people in the nonprofit sector you won't know how well this fits with your views.

For example, people in nonprofits often have a willingness to tolerate ambiguity and inefficiency to a greater extent than you might find in for-profit or military worlds. Most nonprofit organizations have very autonomous workers, even at lower levels of the agency. It's hard to get them to "take orders" without a great deal of conversation and persuasion. Also, client needs and emergencies tend to take precedence over all other tasks. As a grant-writer needing information for a grant proposal that is quickly coming due, you may be amazed at how difficult it is to impart your sense of urgency. Also, on a more charitable note, most nonprofits have staffs that are stretched very thin. They really can't do more than what they are currently tasked with achieving.

Second, you should have a belief in a cause and show that you have supported it in the past. Grantwriting for a nonprofit is often a labor of love that you can get paid for as well. But you need to show that you can truly care about some cause beyond yourself if you want a nonprofit to hire you to support their cause. Your cause doesn't have to be the same as the organization's cause—you don't have to have volunteered for Big Brothers/ Big Sisters to work for them, but the person doing the hiring is interested in what you find important.

Once these two things are part of your background, you can then take steps that are directly related to getting a grantwriting job. Obviously, the more you can show that your education has prepared you for such a position, the better a case you can make. But this is not essential.

What is essential is ***experience***. Experience can be achieved by writing a grant proposal as part of a class (formal or informal); volunteering to help an organization with a grant, getting mentoring from an experienced grantwriter, or, in the worst case, just doing one on your own for practice. You will truly be amazed at how much different it is to actually *do* a grant proposal than it is to *read about* doing a grant proposal. Things that seem clear in the reading are much more complex in the doing.

BEING IN A CLASS OR TRAINING WORKSHOP

Being in a formal class or training has several benefits to you. First, it is something that you can put on your resume and discuss when you begin to look for work as a grantwriter. Classes usually have a clear structure that is helpful in understanding a complex undertaking such as writing a grant. Plus, because there is an instructor, you have someone to ask questions. In the end you frequently have an actual product to show as well as a clearer understanding of the process of writing a grant.

Traditionally classes were face-to-face but now many are available as an online experience. You should still look for a real person you can direct questions to and a way to receive feedback on your efforts. Without these in place, you may find that there isn't that much value to the class.

VOLUNTEERING TO HELP AN ORGANIZATION WRITE A GRANT

This approach is best if done for an organization that you already have contact with as a worker or volunteer in another capacity. You really need to know someone in the agency if you hope to have that person's help. At first, depending on your skill levels, you actually may be more hindrance than help to the grantwriter on staff. It takes a lot of time to teach someone in a mentoring way, which is why my experience starting off as an intern is not an unusual one. There is a great deal of "sink or swim" mentality involved. This is not due to a lack of kindness but a lack of time. You will most likely be expected to work a lot on background research before being put in charge of drafting a section or two. Particularly near the end of the process, when time is tight, the person you're working for may push you off to the side because it will be easier to do the job herself than to let you have a stab at it first.

After it's over, you'll have a bona fide grantwriting experience in an area that is of interest to you. If you've done a good job, you'll probably get the opportunity to do it again! Again, you'll have a real product to show for your efforts and a clear understanding of the process. In fact, because this experience will have been gained in real time with real deadlines, you can really feel the stress and difficulties of the job, as well as the elation for having gotten the proposal turned in.

BEING MENTORED BY AN EXPERIENCED GRANTWRITER

This approach certainly overlaps with the previous method but the difference here is that the mentor is a freelance grantwriter. There is thus no organizational structure in place in the same way that there is volunteering inside an established agency. The trick here is to find an experienced grantwriter who is willing to take you on as a mentee. Most freelancers feel that their time is truly their money, as they have only so many hours in a week to work. Any time spent with a neophyte grantwriter is time that might be spent earning fees. So you need to be very clear what the expectations are for supervision from her/him, and it should also be very clear what you will do in exchange.

One way to find such a mentor is to go to local members of any of the grantwriters' associations that exist. Three examples are:

- American Grant Writers' Association: http://www.agwa.us/
- Association of Fundraising Professionals: http://www.afpnet.org/
- Grant Professionals Association: http://grantprofessionals.org/

There are also groups dedicated to grantwriting on LinkedIn and Facebook that you may use as sources for mentors. While this may be a controversial position, I think a mentee should be willing to pay the mentor to partially compensate for the time spent in training. I certainly would not expect this to be thousands of dollars, but to be included in the entire process from start to finish is a valuable experience for the mentee. All conditions of this sort should be spelled out in advance in order to keep the situation clear.

WRITING A GRANT ON YOUR OWN

I'll be the first to admit that this is a tough way to go. Still, you can buy a book and work through the process on your own. You can also watch YouTube videos that cover grantwriting processes. Websites with free information are available by the scores, including ones put together by federal government agencies that give out many millions of dollars of grants per year.

You will not learn much just from reading a book, watching a video, or looking at websites, but you can learn a great deal if you combine those activities with putting the concepts into practice. If you have the gumption to do this, then you probably have the drive to take your finished product to a grantwriting professional and offer that person the opportunity to review your proposal for a fee. It may be that you can develop a relationship with that person who could then become a mentor. The most important parts of this approach are to do the entire grant from start to finish and then to get feedback. Without the feedback you won't get the full benefits, although what you learn by going through the process will be quite valuable by itself.

GRANTWRITING: EMPLOYEE OR FREELANCER?

Naturally, at some point, you are going to want to be paid for your work as a grantwriter. You'll need to decide whether you want to be an employee of a nonprofit organization (full- or part-time) or be a freelance grantwriter. Let's look at the plusses and minuses of both approaches.

EMPLOYEE

Employees are people who work for one organization at a time. They receive a salary and benefits from that organization. According to the results of my recent survey of grantwriters, over half (57%) of the respondents are employees. You'll work a regular 40-hour-per-week position, but you may not work as a grantwriter the entire time. Many people who work in an agency have multiple hats to wear and may do other types of fundraising work. In fact, many executive directors are the grantwriters in their organization!

This brings up the point that being a full-time grantwriter is not the usual role of someone who writes grants in a typical nonprofit. This type of position is usually only available in very large nonprofits, and mainly to people with already established track records.

Being an employee of a nonprofit, especially as a grantwriter, can be a stressful job. You are responsible for keeping the money flowing. If you aren't successful, other people may lose their jobs. You don't often get to relax, especially if grantwriting is just one of the things you do. You may be forced to write "uninteresting" grants just to keep the doors of your organization open.

Pay and benefits are usually negotiated within a fairly narrow range, as is true of most positions in the nonprofit world. The best way to get a bump in pay is to change employers. We examine pay rates for grantwriters in chapter 2, and found that the median salary for full-time workers is around $65,000 per year.

FREELANCER/CONSULTANT

In medieval times, knights who were not beholden to any particular lord could hire themselves out to whoever needed a warrior. A freelance grantwriter, in a similar way, may write grant proposals for anyone. Freelancing is good if you want to do only work associated with writing grants, so you can set your own terms. It is also good for anyone who does not necessarily wish to have a full-time job. Flexibility and being able to choose what project you want to work on are the major advantages of this type of work.

The disadvantages of being a freelancer include not having an assured source of income. As self-employed people, consultants must pay their own social security taxes, medical insurance premiums, and retirement contributions. They also must negotiate contracts and send invoices to the organizations that hire them. No sick time or vacation is paid for by an employer; when you work as a freelancer, time not working is time not earning income.

According to www.payscale.com, (2013), "the standard rates billed to clients for grantwriting and related consulting services range from a low of $50 per hour to a high of $200 per hour. Most rates range between $65 and $80 per hour." An hourly rate is not necessarily a valid measure, however, of potential earnings. Many consultants do not charge by the hour but bill a flat rate per job. The rate depends on many factors, including the type of funder, the amount of lead time given, the complexity of the proposal, the number of stakeholders to be involved and so on. The flat fees charged by consultants vary from a low of only a few hundred dollars to $15,000 and beyond.

CONCLUSION

The answer to the question "How do I get started as a grantwriter?" doesn't have a clear answer. People get into grantwriting in different ways. One person may find the job thrust on her with no warning. Another may plan out a career trajectory that includes various courses in program design, grantwriting, budgeting, nonprofit management, and more, all to be ready to write grants.

I did a survey of grantwriters recently, asking what the joys of the profession are. While answers varied, four main themes emerged:

- **I love the work**—writing is a creative process that I truly enjoy.
- **I want to support my organization's mission**—grantwriting allows me to contribute to the cause in a big way.
- **I like winning**—getting a grant is a thrill all by itself.

- **I feel valued**—my organization values the work I do as I help it continue to provide services.

If this is the type of motivation that gets you out of bed every day, you should seriously consider the job of grantwriter. The main issue is how to get that first opportunity. If we summarize the steps, they are:

1. Make sure you're a good fit for the type of work this is.
2. Ensure you are competent in the basic skills of writing, being organized, attending to details, having the social and administrative skills to lead groups, and so on.
3. Learn the grantwriting specific skills of finding funding opportunities, understanding what's asked for in requests for proposals (including the jargon that's used), and the mechanisms of proposal transmittal.
4. Get experience however you can.
5. Confidently step forward and call yourself a *grantwriter.*

But don't forget the most important steps after that: keep up to date on your skills, learn new things, and get better at what you already do.

PRACTICE WHAT YOU'VE LEARNED

STEP 1

Look at the list of skills shown below in the grid. Rank YOURSELF on where you are RIGHT NOW in terms of your level of that skill.

	My Current Level of Competence				
Skill	**I don't have this skill at all (0)**	**I have very little of this skill (1)**	**I can just barely get by with what I currently know (2)**	**I'm pretty good at this but can see the need to get better (3)**	**I am really quite skillful in this area (4)**
Technical writing					
Organization					
Attention to detail					
Group leadership					
Following directions					
Finding funding opportunities					
Reading requests for proposals for important elements					
Researching social problems for specific geographic areas					
Finding evidence-based programs					
Developing and/or modifying programs to address social problems					
Assessing organizational capacity					
Designing evaluations					

STEP 2

For each of the skills you have assessed yourself at a 0, 1, or 2 level, write a paragraph on how this impacts your desire to become a grantwriter. Prioritize the skills you feel you need most. Then, for the topmost three skills, develop a detailed plan to get yourself to at least a 3 level in the next three months. What resources will you need to access? How will you acquire that access? Whom do you know or whom can you pay to help you with getting up to a 3 level?

#1 MOST IMPORTANT SKILL TO IMPROVE: ____________________

Plan to Improve

List resources needed to move to at least a level 3

____

____

____

____

____

____

Timeline to Improve

List specific milestones you will reach and when you will achieve them to move to at least a level 3 for this skill. Make milestones small so you will be able to achieve them and gain or keep momentum: for example, say "I will read one chapter in a book about grantwriting on this topic each week for two weeks" rather than "I will find how to apply this skill to five grant proposals by the end of the tomorrow." Keep your milestones close together in terms of small skill improvements and also in terms of amount of time needed to accomplish.

____

____

____

____

#2 MOST IMPORTANT SKILL TO IMPROVE: ____________________

Plan to Improve

List resources needed to move to at least a level 3

Timeline to Improve

List specific milestones you will reach and when you will achieve them to move to at least a level 3 for this skill. Make milestones small so you will be able to achieve them and gain or keep momentum: for example, say "I will read one chapter in a book about grantwriting on this topic each week for two weeks" rather than "I will find how to apply this skill to five grant proposals by the end of the tomorrow." Keep your milestones close together in terms of small skill improvements and also in terms of amount of time needed to accomplish.

#3 MOST IMPORTANT SKILL TO IMPROVE: ____________________

Plan to Improve

List resources needed to move to at least a level 3

Timeline to Improve

List specific milestones you will reach and when you will achieve them to move to at least a level 3 for this skill. Make milestones small so you will be able to achieve them and gain or keep momentum: for example, say "I will read one chapter in a book about grantwriting on this topic each week for two weeks" rather than "I will find how to apply this skill to five grant proposals by the end of the tomorrow." Keep your milestones close together in terms of small skill improvements and also in terms of amount of time needed to accomplish.

STEP 3

Look at each of the skills you have stated you are at a 3 level. Which of these skills do you absolutely want to move to a level 4 status? How will this help you, specifically? Then, write a paragraph describing how you might move to a level of 4.

#1 MOST IMPORTANT SKILL TO IMPROVE FROM A 3 TO A 4: ______

Plan to Improve

List resources needed to move from a level 3 to a level 4

Timeline to Improve

List specific milestones you will reach and when you will achieve them to move to a level 4 for this skill. Make milestones small and timeline tight so you will be able to achieve them and gain or keep momentum as you did for your lower level skills.

#2 MOST IMPORTANT SKILL TO IMPROVE FROM A 3 TO A 4: _____

Plan to Improve

List resources needed to move from a level 3 to a level 4

Timeline to Improve

List specific milestones you will reach and when you will achieve them to move to a level 4 for this skill. Make milestones small and timeline tight so you will be able to achieve them and gain or keep momentum as you did for your lower level skills.

CHAPTER 2

GRANTWRITING IN THE AGE OF SCARCITY

This chapter describes the grantwriting world of today, the Age of Scarcity. The job of grantwriting has never been easy, but it is certainly more difficult now than it has been for decades. That's because we are living in a quickly changing funding environment. To truly understand grantwriting, we must understand how it fits into the larger world of nonprofit organizations and how they acquire resources.

Grantwriting by nonprofits is the act of developing a proposal to be sent to a funder (which could be a government agency, a foundation, a corporation, or other organization) requesting support for a program or services to benefit society. The most common sections of any grant proposal include description of a problem, a plan to solve the problem with specific goals, a way to ensure that the solution is implemented according to plan, and a budget for achieving the promised goals. Proposals often include a way to evaluate the effectiveness of the project as well as other sections required by the funder.

THE BROADER CONTEXT OF GRANTWRITING: THE AGE OF SCARCITY

Grantwriting needs to be seen in its broader context, the world of fundraising for nonprofit organizations. In general, human services nonprofits acquire resources in three main ways: individual donations, grants from foundations, and grants from government agencies. Other sources of support do exist, such as corporate donations, special events, and social enterprise efforts, but these are relatively small in terms of percentage of an organization's budget. These three main legs of nonprofit funding have become less stable in recent years, and often less abundant as well. Because nonprofits have few ways to acquire funds, when instability rises and competition for resources increases, they face difficulties in planning and providing adequate services for their intended recipients.

The Nonprofit Finance Fund (2015) reports many worrisome facts about what is happening to nonprofits and their clients in the Age of Scarcity:

- 76% of the nearly 5,500 nonprofit respondents said that demand for services increased in 2014, compared to 2013. This means demand for services increased for seven years in a row.
- Over half (52%) of all nonprofits were not able to meet the demand for their services in 2014. This is the third consecutive year that more nonprofits could not meet demand than could.
- Nonprofits in the human services arena were more hard-pressed than the collective group: 61% of them could not meet demand.
- When client needs overwhelmed agency capacity, 71% said that clients were not able to get their needs met elsewhere.

Even as the economy has improved, compared to the worst days of the Great Recession, which lasted from December 2007 to June 2009, Antony Bugg-Levine, the leader of the Nonprofit Finance Fund stated, "The social sector is in flux, with many organizations moving beyond 'crisis mode' but still facing an uncertain future" (Nonprofit Finance Fund, 2015). The economic situation nonprofits face today is precarious, with even long-established and formerly successful human service providers ceasing operations due to financial problems.

THE THREE MAIN LEGS OF NONPROFIT FUNDRAISING

The number of nonprofits registered with the US Internal Revenue Service grew by 24% between 2000 and 2010 (University of San Francisco, 2013). This startling growth in nonprofit organizations has led to greater competition for funding of all types. Despite there being a number of ways that nonprofits could raise funds, the typical nonprofit has three legs of funding, even if they don't use all of them consistently or well. These are individual giving, foundation giving, and government grants and contracts. Let's examine the current levels of funding for each of these sources.

INDIVIDUAL GIVING

For a few years during the Great Recession, individual giving decreased considerably across the nonprofit world. Fortunately, in 2013 the tide turned and overall charitable giving increased by 4.9% (Blackbaud, 2014). Human services organizations had a modest 3.6% increase in 2014 compared to 2013. This positive trend for the nonprofit sector as a whole continued in 2015, although it slowed to an overall smaller giving increase of 1.6%. The subsectors of international affairs and faith-based organizations grew the most between 2014 and 2015 (5.1% and 3.9%, respectively). For two subsectors, however, the story was grim. Public and society organizations saw a decrease of 0.9% in giving, and, most to the point for this analysis, donations for the human services subsector dropped by 2.8% in 2015 compared to 2014 (Blackbaud, 2016).

In short, human services nonprofits faced higher demand for services but had fewer resources from individual donors to assist those in need. There is hope that the US economy will improve and donors will once again increase their giving.

The problem for nonprofits is that no one can be sure just how well the economy is going to do in the next few years. Some economists argue that there will be a stock market correction with a resulting loss of stock portfolio value, causing individuals to again lose significant ground financially. If this occurs, they will be less likely to give, just as they gave less during the Great Recession. Already, many baby boomers are working more years because they see that their retirement savings are not as large as needed. They are also seeing that they may need to continue assisting their children get started in their jobs and in raising their own families. This may bring forth a "charity-begins-at-home" mindset, also leading to less generous giving to nonprofit organizations serving others.

In addition, income inequality is continuing to grow, with the percentage of Americans in the middle class declining. "U.S. income inequality has been increasing steadily since the 1970s, and now has reached levels not seen since 1928," according to the Pew Research Center in a blog post on December 5, 2013. This assessment is backed up by the latest figures from the Bureau of Labor Statistics (Karageorge, 2015). Thus, it is not clear what will happen with individual donor giving. A final drag on individual giving's value for nonprofits is that it is likely that inflation will increase in coming years because it is at historically low levels now. This means that even if more individual donations are given, they will be worth less to organizations. Scarcity in individual giving for human services organizations will likely continue to be problematic.

FOUNDATION FUNDING

According to the Foundation Center (2014), there were over 84,000 foundations in the United States in 2014. They are responsible for 15% of all private giving in the United States (Blackbaud, 2016) and gave approximately $22.4 billion away in 2013 (Foundation Center). Human services organizations received about 16% of total foundation giving, or $3.5 billion. Giving by foundations increased in the past few years after falling to a recent low in 2010. Still, once inflation is factored in, the increases are quite small, less than a 1% increase. This is not enough to keep up with the greater demands for services, which have increased for seven consecutive years (Nonprofit Finance Fund, 2015). Foundations are quite dependent on their stock market portfolios to be able to give to nonprofits. If (or when) the stock market loses value, foundations will have fewer resources to give to applicants, again squeezing nonprofits just as need increases.

GOVERNMENT GRANTS AND CONTRACTS

The impact of the Age of Scarcity is especially prevalent in the government sector. When government grants and contracts decline, the nonprofit sector feels the pinch quickly because of the extent of government funding for nonprofits. For example, in 2011, government grants and contracts provided one-third of revenue for public charities (Pettijohn & Boris, 2013). Results from a survey of human service nonprofits showed that, in 2012, "almost 30,000 nonprofits reported close to $81 billion in government contracts and grants" (Pettijohn & Boris, 2013, p. 35). In 2013, nearly 50% of organizations surveyed indicated they experienced a decrease in local, state, or federal funds (Pettijohn & Boris).

The Great Recession caused many problems for nonprofits, most of which were still problems when Pettijohn and Boris published their data in 2013, several years after an economic recovery began. One major federal government response to lasting budgetary problems occurred on January 2, 2013. This is the day when the US government went over the "fiscal cliff" and automatic budget cuts (sequestration) were triggered. Although cuts are being phased in over several years, federal government funding for many human services programs and the nonprofits that actually run them have been hit hard and continue to be the targets of budget cutters. In fiscal year 2013, sequestration cuts at the federal level were about $85 billion. This was about 5% of that year's budget, and cuts were applied equally across all programs (Health and Human Services, 2013). The national government still provides tens of billions of dollars of funding per year but, as with other sources of income, competition is greater and the need to be an excellent grantwriter is heightened if a nonprofit is relying on the national government for support.

States vary tremendously in their political support for human services programs and thus the level of state money allocated to help vulnerable populations and the nonprofits that serve them. Some states, such as Connecticut, have shifted funding in recent years away from education and human services to employee healthcare and debt (Fiscal Policy Center, 2013). In many states, nonprofits that have state government contracts are just not being paid or they are being paid 4 or 6 months late. State and local governments are delaying reimbursing nonprofits, and some nonprofits are closing down, firing people, or furloughing staff.

States and localities, with problems of their own, are generally not providing enough funding to make up for cuts elsewhere. Much of the funding for human services at the state and local levels comes from pass-through grants from the federal government, and, as we have seen, those funds are not as widely available to states and localities as they once were. The impacts of the Great Recession have begun to fade in some states, but in many austerity remains. Political experiments in deeply decreasing taxes at the state level have generally led to larger government budget deficits, straining nonprofit budgets, sometimes beyond the breaking point.

With this information on funding for nonprofits as background, let us turn to some of the basic facts about the world of grantwriting. Let's start by delving more deeply into what the differences are between foundation and government funding.

UNDERSTANDING DIFFERENCES BETWEEN FOUNDATION AND (FEDERAL) GOVERNMENT GRANTS

In many ways foundation and government grants are similar in nature, in that the funders require you to provide them with an application containing requested information. But the two types of funders are quite different in other ways (see Table 2.1). This section looks at eight specific areas of difference that are important for grantwriters to understand.

Foundations are almost always funded themselves with private money, although the donations to foundations are tax-advantaged for the donors. Thus, in some sense, even foundations are funded with public money because the donors reduce their taxes, thus

TABLE 2.1: Comparing Government and Foundations as Grant-Makers

Aspect	Foundation	Government
Source of money	Private money, though it is usually tax advantaged	Taxes (public)
How announced	Through announcements decided by foundations, usually on their websites, mailings (electronic and regular), Facebook pages, and blogs	Public announcement in official sources. Federal: grants.gov State/local: varies by jurisdiction
How submitted	Varies: may be electronically or paper	Electronically (very rare exceptions)
Specifics described in funding opportunity announcements	Usually general aims of foundation are described with less specificity than government RFPs	Quite detailed with room for varying creativity and rigor
Decision-making process	Screened for match with foundation criteria (both explicit and implicit). If okayed, project officer reviews. May ask for additional information. Final decisions tend to be made by foundation board of directors. Some chief operating officers may be able to make decisions on all or some applications.	Screened for completeness then reviewed by panel of experts, who score all applications. Highest-scored applications receive further scrutiny and possibly are funded.
Length of grant	Frequently 1-year, but varies	Frequently multiyear, but varies; even multiyear have to be reapplied for.
Post grant award	Amount of oversight varies considerably but generally less than for government grants	Considerable oversight, both financial and programmatic
Connection with other grantees	Unusual to have formal connections with other grantees	Often considerable with annual meetings with other grant recipients of same RFPs

taking money away from government revenues. Government funding at all levels comes directly from tax revenues and is thus dependent entirely on the economic and tax systems at work that fund the government at large.

Grant funding opportunities are announced by foundations through posting on web sites and shared via regular mail or e-mail to selected individuals (or whoever is on their e-mail list). Recently, Facebook pages and blogs are ways to spread the word of funding availability as well. Government grant announcements come via official sources. The federal government uses www.grants.gov to post all funding opportunities. States and local jurisdictions have numerous avenues that grantwriters must become familiar with.

Grant applications are submitted either electronically or on paper for foundations. Government applications are almost always submitted electronically (at least at the federal and state levels), usually through special websites set up for this purpose.

When the funding opportunities are described, foundations usually stick with the general aims of the foundation and ask applicants to match those aims. Government funding opportunities are usually quite detailed in providing a clear framework for what is required of applicants, while at the same time providing room for creativity on the part of applicants in how to place their ideas within that framework. A government funding opportunity announcement (FOA) is almost always going to be longer and more prescriptive in what is required than is a foundation grant requests for proposals (RFPs).

The decision-making process within foundations can vary significantly. Sometimes the decision is made through negotiations between an agency and the foundation even before the proposal is submitted. Usually, however, all proposals are first screened for a match between the proposal and the foundation's criteria for funding. If the proposal passes this stage, additional information may be asked for to flesh out a letter of inquiry or short proposal. Final decisions tend to be made by the foundation's board of directors. Sometimes the foundation's chief operating officers may be able to award some grants on their own. Government grants follow a structured review process in order to assure as much impartiality as possible. Independent reviewers are selected to review and score applications. The applications with the highest scores receive further scrutiny and are possibly funded, depending on the amount of money available and other factors, such as geographic dispersion of the awards. It is not true that the highest rated applications automatically are funded.

Foundation awards tend to be 1 year in length, although this can vary from time to time and foundation to foundation. Government grants are frequently 3 to 5 years in length, but with the necessity of reapplying each year in order to demonstrate adequate progress and sound financial practices.

Once an award is made, foundations provide varying amounts of oversight. Some foundations require quarterly reports and clear financial reporting, while others request only an annual report. Government agencies provide a very high level of oversight, including progress and fiscal reports after 6 months and annually, and sometimes site visits by program officers to check up on the grant recipients.

The final area of differences relates to connections with other grantees. With foundation grants, it is unusual for recipients to have formal connections with other grantees. With government grants, agencies often have considerable interaction with other grantees at annual meetings, phone conference calls, website discussions, and in other ways.

WHAT ARE THE BIGGEST MYTHS ABOUT GRANTWRITING?

When you speak with people who aren't grantwriters, you may hear a number of myths about the field. At least one federal grant-giving agency has a page on its website (Health Resources and Services Administration, http://www.hrsa.gov/grantmyths/) about the myths of grantwriting that it has to deal with. Here we cover five common myths.

MYTH 1: GETTING A GRANT IS NEARLY IMPOSSIBLE

Competition *is* tough for some grants, but in the end someone is always chosen. With the right skills and efforts, there is no reason why your proposal shouldn't be among those selected. It's a myth that getting a grant is impossible, but it should be understood that considerable work is involved. Also, there are degrees of difficulty. Large federal government grants are probably the most difficult to be awarded. If you're just starting out, these are NOT the place to start, if you can help it. At the same time, foundations must award 5% of their assets every year, so if you do your homework, select foundations that have an interest in the work you propose, and craft a good proposal, you can definitely increase the odds of your success.

MYTH 2: GETTING A GRANT IS EASY: SOMEONE HAS TO GET THE MONEY, RIGHT?

This is the flip side of Myth 1 and it is equally incorrect. Despite what sellers of grantwriting trainings and books may say, writing a strong proposal requires a lot of skill and knowledge. Competition for different grants can be quite stiff. Even strong proposals may be turned down for reasons that have nothing to do with the proposal itself. The writer must attend to a myriad of details, all of which are important. Losing a few points due to carelessness or poor understanding of the intentions of the funder may be the difference between being funded or being rejected for support.

MYTH 3: A SUCCESSFUL GRANT IS JUST ABOUT FILLING IN THE BOXES CORRECTLY

A substantial part of the success for any grant proposal is following directions. But the "boxes" need to be filled in with well-researched problem statements and creative ideas regarding how to solve those problems. On top of that, the proposal must include clear plans to get the program off the ground. The implementation of the solution must be monitored and the outcomes evaluated. In addition, a great deal of infrastructure needs to be in place and documented before the funder will believe you are capable of handling the administration of the grant. As you can see, a lot goes into a grant proposal. The grantwriter can influence some aspects of the process—much is beyond even the best proposal author's control.

To consistently write good proposals, grantwriters need to have a great deal of preparation and lead time. While it's neither impossible nor easy to get a grant, it requires someone with a considerable amount of knowledge and expertise to put together all the details that go into a successful grant. Excellence (and success) in grantwriting is based on general excellence in staffing, organization, and leadership. It's not JUST about filling in boxes or blanks on an application form.

MYTH 4: GRANTWRITING IS A HIGHLY LUCRATIVE FIELD

Two different grantwriting fields exist: freelance and working for a single organization. Both will let you earn a good living, but won't necessarily get you beyond a solid middle-class lifestyle. Freelance grantwriters can bring in a very good income, though they work hard and the pay is episodic. You have to be among the best in your field, with an excellent track record, before you can command the high fees that are sometimes bandied about as if they are average compensation levels. These grantwriters can earn $10,000 or more per grant, whether the proposal is chosen for funding or not. For people with the job title of grantwriter working for an organization, the median salary in May 2016 was $64,396 per year, with virtually all grantwriters making between $58,000 and $72,000 per year (http://www1.salary.com/Grants-Proposal-Writer-Salary.html).

MYTH 5: GRANTWRITERS ARE PAID A PERCENTAGE OF THE FUNDS THEY BRING IN

A myth related to compensation is that grantwriters are paid a percentage of the funds they bring in. The truth is that is usually considered unethical. The Code of Ethics of the Association for Fundraising Professionals explicitly says that grantwriters are to be paid whether the proposal is funded or not. Think of it this way—doctors get paid whether the patient gets well or not; lawyers get paid whether the case is won or not—and grantwriters should also be paid whether they are successful or not. While there may be a few grantwriters willing to take a percentage of a successful grant as their compensation, it is generally looked down on by professionals in the field. Almost all proposal writers believe there are too many factors determining whether a particular grant is funded or not to have their income dependent on the vagaries of chance.

TWO IMPORTANT TRUTHS

Now that we've explored five myths about grantwriting, we should look at two important truths for both foundation and government grants. First, finding appropriate grant sources takes a lot of time and, two, you have to develop and use your detective (research) skills if you're going to be a capable writer of grant proposals.

FINDING GRANT SOURCES TAKES A LOT OF TIME

One reason that a lot of time is required is because it takes a long time to do the research to find potential sources of funding. We'll go over a number of systematic approaches you can use to find funding in chapters 3 and 4, but at this point we can say that you can't just expect to type in one key phrase into Google or any other search engine and find a lot of good potential funding sources. There are many reasons why finding appropriate grants is not for the faint of heart. This relates to the way funding is structured. The funder that you're interested in may not be offering a grant right then. Grants come in cycles.

Foundations may only make decisions regarding the grant proposals that have come in every 3 or 6 months. So you may just be the victim of poor timing if you put in a proposal shortly after the deadline for the current period passed. You will have to wait until a new deadline comes up where your proposal is considered. It is important to try to turn your proposal in at a time not too much before it would be considered.

With federal government agencies, the cycle for funding fits within fiscal years. The fiscal year for the federal government begins October 1. That means agencies are trying to allocate all of their funding by the end of the September. Up until the Age of Scarcity, what you often saw was that agencies that projected that they are going to have unallocated funding at the end of the fiscal year would start finding ways to use it so that they could say, "We've used all of the funding we were given." Thus, the last quarter of the year tended to have a lot of requests for proposals. (The last quarter of the federal fiscal year is July 1–September 30.) Now, in the Age of Scarcity, with many core services being reduced throughout the year, we probably will see less of this behavior.

The implication for nonprofits seeking funding from the federal government is thus that you must be very aware and keep close tabs on the RFPs that do come out. You won't be able to count on additional funding opportunities to suddenly come up in late summer and early fall.

State government grants have a different fiscal year, which can vary from one state to another. A large number of states start their fiscal year on July 1 and end it on June 30. Just as with the federal government, with state government budgets under considerable stress, discretionary grant announcements are unlikely to be as plentiful in the fourth quarter, from April 1 to June 30, as they used to be.

Government funding is likely to be much more attached to a regular schedule, with funding announcements for particular topic areas coming out at the same time every year. As a grant writer you must understand what that cycle is. It's going to be different for different agencies.

The reason it takes a lot of time to find grant opportunities (both foundation and government), is because you have to find the opportunities and then you have to wait for the opportunities to be open for submissions. At that time, when the announcement comes out, you'll begin the actual writing and fine-tuning of the grant. (Actually, as explained later, you shouldn't wait to get started until the time the RFP is released, but you will have to fine-tune what you write in response to what is in the actual proposal request.)

Writing a grant *always* takes longer than you think it will. Even when you get to be an accomplished grantwriter, you will take longer than you first believe you will. It seems there is always some little detail that you're missing. Also, grant applications must be signed by an authorized individual, so you need that signature and it may not be possible to get immediately. In many agencies, the board of directors has to approve the submission of the proposal, and it may not be meeting for a while. These are all reasons why it takes longer to finish the grant than you expect. The proposal, when completed, will then be submitted and reviewed, with finally a decision made. The review and decision-making process can take a number of months all by itself. There is really nothing you can do to speed the process up.

YOU HAVE TO DEVELOP AND USE YOUR DETECTIVE (RESEARCH) SKILLS

In my experience, few people truly enjoy their mandatory "research" classes they take in college. I think this is because research courses are presented as a set of unfathomable rules for doing pointless studies about narrow topics of interest to only a very small number of people in the entire world. If I could do one thing to improve the college experience for everyone, I would rename "research" courses as "detective" courses. This brings about a completely different image. Instead of ivory-tower professors pursuing tiny bits of knowledge, students would imagine themselves as following in the footsteps of Sherlock Holmes to solve crimes based on careful collection and weighing of information that leads to a conclusion as to who committed the deed.

Grantwriters need such careful information collection at many different points in the process, but here we concentrate on finding appropriate funding opportunities. Sleuthing out where the most appropriate grant requests are and what the foundation or government agency is asking for is the type of research you need to do. You need to become like the detective trying to find the criminal by thinking like the criminal, but in this case, you are trying to achieve the feat of getting a grant by thinking like a funder. This requires detective (research) skills but is actually not as difficult as it may seem at first.

HOW DO YOU FIND OUT WHAT FUNDERS WANT TO GIVE MONEY TO?

The last topic in this chapter links well with everything that has come before. The Age of Scarcity makes the job of a grantwriter more important than ever. Being a successful grantwriter when resources are tight is vital. One aspect of the grantwriting job that we home in on in later chapters is the importance of matching what funders want to give their resources for with what you propose. If you can do this, you increase your ability to write winning proposals. The question is how to know what funders are willing to support.

Funders are actually happy to tell you what they are looking for—they don't want to waste their time by reading proposals that don't match their interests. But all too often grant writers don't look for or collect information on what funders want in a thorough way. This results in wasted time, unfunded grant proposals, and significant levels of frustration.

Foundations frequently post on their website the types of ideas that they want to give attention (and money) to. You can often read at least synopses of recently awarded grants, which provide you with hard data on the decision-making outcomes of that foundation. Government agencies also describe through position papers and strategic plans what their goals and objectives are for the future, and describe what issues are important for them. Grantwriters need to look for such information and then read it carefully to understand what to focus on for future funding opportunities from those government bodies.

In short, the top proposal writers learn what potential funders want to support and they begin to gather information to support proposals for programs, interventions, and services that match what the funders want to give money for. Grantwriters need to use their detective (research) skills in this way to find funding opportunities that are highly targeted to match the capabilities of the nonprofit they are working for.

CONCLUSION

This introduction to the world of grantwriting in the Age of Scarcity emphasizes a couple of points. First, this is both a difficult time to be fundraising for nonprofits and perhaps the most important job if you want to support an organization or cause. Second, foundation and government grants, while similar, have significant differences. Third, despite these differences, whether you're looking for a foundation or a government grant, you must allocate a lot of time in finding appropriate funding sources. Chapters 3 and 4 provide more detail on this, but if you don't allow enough time, you're not likely to overcome the increasingly difficult odds of writing a successful grant.

You'll have to put on your detective hat and use your research skills to find appropriate sources of funding. Research skills are needed in other parts of the grantwriting process, but you'll need to start off with them. One of the most important research-related tasks, one that will help you think like a funder, is to find and study the funding priorities of all the foundations or government agencies that you consider as likely places to submit a grant application to.

In the chapters that follow, you will be able to learn and practice what you need to know to become a grantwriter. After each chapter you'll be given a chance to practice what you've learned. If you skip these skill-building exercises, you will be short-cutting the learning process and prolonging becoming a professional grantwriter. If you *are* willing, then even if you find some things difficult or challenging, in the end, you will succeed—you'll have written a grant proposal that you can submit. So, you need to decide: "Are you ready and willing?" The material is here—it's up to you.

The next two chapters focus on how you most efficiently find the grant opportunities that remain in the Age of Scarcity. Chapter 3 shows you how to uncover foundation grant opportunities and Chapter 4 shows you how to find government grants.

PRACTICE WHAT YOU'VE LEARNED

1. Talk with fundraisers at local nonprofit organizations about the ways that the Age of Scarcity has affected their organization, or do some research on any nonprofit you want to consider. How has their fundraising changed over the past decade? Are their grantwriting efforts more or less successful? What other information can you uncover about impacts in your area or subsector?
2. Begin conversations with people you know, even colleagues with experience in the nonprofit sector, about what they "know" about grantwriting. Do they believe any of the myths you've read about in this chapter? How accurate is their impression of the field, based on what you know at this point?
3. Find a few foundation websites and look them over. Just from that experience, can you describe what they might be interested in providing resources for?

REFERENCES

Blackbaud (2014). Charitable Giving Report: How Nonprofit Fundraising Performed in 2013. Retrieved from https://www.blackbaud.com/files/resources/downloads/2014/2013.CharitableGivingReport.pdf

Blackbaud (2016). Charitable Giving Report: How Nonprofit Fundraising Performed in 2015. Retrieved from https://www.blackbaudhq.com/corpmar/cgr/how-nonprofit-fundraising-performed-in-2015.pdf

Fiscal Policy Center (2013). Report: State Funding Mix Has Shifted from Education and Human Services to State Employee Health Care and Debt. Retrieved from http://www.ctvoices.org/sites/default/files/bud-12shiftingprioritiesrel.pdf

Foundation Center (2014). Key facts on U.S. Foundations. Retrieved from http://www.foundationcenter.org/gainknowledge/research/keyfacts2014/foundation-focus.html

Health and Human Services (2013, May 17). CSBG Dear Colleague Letter Sequestration Update May 17, 2013. Retrieved from http://www.acf.hhs.gov/programs/ocs/resource/csbg-dear-colleague-letter-sequestration-update-may-17-2013

Karageorge, E. (2015, April). The Growth of Income Inequality in the United States. Bureau of Labor Statistics. Retrieved from http://www.bls.gov/opub/mlr/2015/beyond-bls/the-growth-of-income-inequality-in-the-united-states.htm

Nonprofit Finance Fund (2015, April 27). Survey: Need for Housing, Youth Development and Jobs Outpacing Available Services; 52% of U.S. Nonprofits Can't Meet Demand. Retrieved from http://www.nonprofitfinancefund.org/announcements/2015/2015-state-sector-survey-results-release

Pettijohn, S. & Boris, E., with De Vita, C. & Fyffe, S. (2013). Nonprofit-Government Contracts and Grants: Findings from the 2013 National Survey. Urban Institute. Retrieved from http://www.urban.org/sites/default/files/alfresco/publication-pdfs/412962-Nonprofit-Government-Contracts-and-Grants-Findings-from-the-National-Survey.PDF

Pew Research Fund (2013, December 5). US Income Inequality, on Rise For Decades, is at Highest Since 1928. Retrieved from http://www.pewresearch.org/fact-tank/2013/12/05/u-s-income-inequality-on-rise-for-decades-is-now-highest-since-1928/

University of San Francisco (2013). The rise of the nonprofit sector [Infographic]. Cited in Lambert, J., (2013, Dec. 10). Infographic: What is driving nonprofit sector's growth? *Nonprofit Quarterly*, Retrieved from https://nonprofitquarterly.org/2013/12/10/infographic-what-is-driving-nonprofit-industry-growth/

CHAPTER 3

FINDING FOUNDATION FUNDING SOURCES

This chapter examines two primary ways to find information on foundations that may be good matches for your organization: using the Foundation Center's online directory, and conducting searches on the Internet. Both are useful alternatives, although you may find you can't access the Foundation Center information at a price you want to pay. Thus, you should know some of the tricks to searching on your own, using Google, Bing, or other search engines.

Before covering this material, it is important to discuss briefly what foundations are and the varieties of foundations that exist in the United States. Without knowing this information, it is difficult to swim in the waters of the foundation world. We will also cover how the Age of Scarcity has impacted foundations of all types.

WHAT ARE FOUNDATIONS?

According to the Minnesota Council on Foundations (MCF), "A foundation is a nonprofit organization that supports charitable activities in order to serve the common good" (MCF, 2014). According to the Foundation Center (2014), in 2012 there were 86,192 foundations in the United States. They held $715 billion in assets and gave away $52 billion.

There are several different types of foundations that exist under the law. The first, independent foundations, are the most common and typically are created by individuals or families that want to promote attention to a certain problem or approach to a problem. Independent foundations can be either family foundations or other independent foundations, although there is no precise legal definition of the term "family foundation," as they are part of the larger category of independent foundation (MCF, 2014).

Perhaps the world's largest foundation is an independent family foundation, the Bill and Melinda Gates Foundation, which gave away $3.4 billion in both 2011 and 2012, and $3.6 billion in 2013 (Gates Foundation, 2014). The key elements demarcating family foundations, according to the MCF, are that the funding comes from members of a single family, and at least one member of the family continues to provide strong leadership for the foundation. Other independent foundations do not start off with or continue with a strong family identify.

Corporate foundations are set up by corporations as legally separate entities, overseen by a board of directors often made up of corporate directors and employees. Funding for corporate foundations varies, but can include an endowment, contributions from current corporate profits, or even donations from employees. Some corporate foundations provide grants only to locales or states where the parent company has a strong presence; others have broader eligibility criteria. Corporate foundations are not the same as corporate giving programs, which often donate goods and services rather than cash, or provide only small amounts of direct funding for nonprofit projects. Another difference is that corporate foundations are governed by Internal Revenue Service rulings and law, while corporate giving programs are entirely in the hands of the corporation that they are a part of.

Community foundations are the third major type of foundation. These are tied very closely to a particular geographic area and are usually funded by pooling smaller amounts of donations from people in the community. Donation decisions are made by a board of directors that is supposed to be representative of the community at large.

All private foundations are required to follow at least three very important regulations in order to maintain their standing as foundations according to the MCF (2014). First, they must pay out (donate) no less than five percent of the value of their investment assets. Second, they pay taxes of one or two percent of their earnings. Third, with rare exceptions, they can only donate to other organizations that are 501c3 (charitable) organizations.

Not all organizations with the word "foundation" in their name provide grants to applicants. The Henry J. Kaiser Family Foundation, for example, uses its resources to develop nonpartisan information on healthcare issues (Henry J. Kaiser Family Foundation, 2014) so it is important when searching for foundation funding, to carefully look up information on each prospective foundation.

HOW HAVE FOUNDATIONS BEEN AFFECTED BY THE AGE OF SCARCITY?

Foundations continue to see ups and downs in their assets, as described in chapter 2. The severe downturn in their fortunes during the Great Recession of 2007 to 2009 has caused them to make some changes in their procedures and practices that are not going to change soon.

One such change is in reaction to what some grantwriters did during the worst of the economic recession. Because grants were increasingly difficult to get, these writers flooded numerous foundations with large numbers of applications, even if the foundations were not a good match for the proposal idea or organization. The rationale of the grantwriters was twofold. First, they might get lucky and have a proposal funded—certainly, a proposal that wasn't submitted wasn't going to get funding, so the more applications that went out, the greater the chance that one of them would strike gold. The second rationale was that they (the grantwriters) only got paid for submitting proposals, not for getting funded (in fact, getting paid a percent of money brought in or based on success is considered an unethical payment approach). Thus, it was vital for grantwriters (freelancers or on staff) to crank out proposals for their own sake, to be seen to be doing their job, even if none of the applications were successful. After all, "everyone knew" it was a tough time to get funding, so grantwriters who sent out slews of poorly targeted or unsuccessful proposals were not seen in the negative light they should have been.

The impact on foundations was sheer overwhelmingness. They received far more applications and on a wider variety of topics than they were prepared to handle. Because of this, and their declining pool of resources, foundations made changes in the way they conducted business. One of the most important shifts was to become much more focused on the problem areas they were prepared to provide funding for. Rather than being open to a fairly unrestricted set of issues, they became much more selective, choosing to have one or a very few topics in which they specialized. Another very important change for some foundations was that they began to look for the nonprofit organizations that were leaders in certain problem areas or communities. Once identified, foundation officials would reach out to those nonprofits to request funding proposals, while at the same time declining to accept unsolicited grant applications.

Both of these changes in foundation practices make a great deal of sense in the context of an Age of Scarcity, with its overly large numbers of applications and fewer resources to donate. What is interesting is that foundation practices are expected to remain this way, even if (when) financial constraints are lessened. Other changes made by foundations reported by the Foundation Center, such as reducing the size of staff, delaying nonessential expenditures, and other "normal" belt-tightening moves made by organizations in times of fiscal stress may be overturned, but the shift to a more narrow programmatic focus, refusal to accept unsolicited proposals, and focusing on reaching out to successful and prominent nonprofits to request applications are not likely to change in the foreseeable future. These impacts of the Age of Scarcity are here to stay.

The next sections of this chapter discuss two major means for doing research on foundations—the Foundation Center online directory and search engines on the Internet, such as Google. Other sources of information exist as well that are more narrowly focused. The MCF, for example, has a database nonprofits can pay to gain access to and there may be such a resource in your state or locality as well. It is simply beyond the scope of this book to gather the full gamut of possibilities together.

THE FOUNDATION CENTER ONLINE DIRECTORY

The number one source of information regarding foundation grants is undoubtedly the Foundation Center (www.foundationcenter.org). According to its website, it has access to over three million funding records and opportunities from 90,000 grant-makers. The major problem with using the Foundation Center's materials is that it is not a free service. At the time of this writing (June 2017), the "Essential Plan" (which provides access to information about more than 100,000 foundations in the United States and other benefits) costs $49.95 per month, or $399 per year. Other plan levels exist as well. The top-of-the-line "professional" level, lets subscribers see information about over 140,000 foundations, review of over 3.8 million grants that have been awarded, and much more for only $199.99 per month (no contract), or $2,098.60 for two years if paid for in one payment, which lowers the cost to just $87.44 per month (Foundation Center, 2016).

While this service can be extremely useful, the cost of access to Foundation Center information is probably outside the price range of most nonprofits that have only an occasional need to actually use this vast array of data. Fortunately, many public libraries carry a subscription at some level and grantwriters can receive training from the library staff to use the Foundation Directory Online (FDO) on the premises. Sometimes access to the FDO is available through other entities, such as a nonprofit consortium, university, or government agency. In all cases, it is important for the person using the directory to receive training on the ins and outs of the system and what is actually going to be useful for the agency to look for. Some grantwriting consulting firms also pay to have access to this information because they need it to assist many different nonprofits. If you hire such an organization, your consultant will have access to the database.

Because of the proprietary and copyrighted nature of the Foundation Center's information, I cannot cover the directory in any more detail than this. If you can get access to training and the information through one of the over 450 local libraries and nonprofit consortia that provide it to the public, you will find it very useful. Online resources exist to help you get the most from it. The Denver, Colorado, Public Library, for example, provides free access and has created a short video explaining how it works. This video is available at http://youtu.be/tKlb8iurAK4 (Denver Public Library, 2013). Techsoup, a nonprofit resource organization, in cooperation with the Foundation Center, has an hour-long video that explains in much greater detail many of the features of the FDO. It is available at http://youtu.be/PzFu13s-6CQ (Techsoup, 2012). Even if you don't use the Foundation Center directory, this video presents a number of very good tips for database searching in general.

I don't think it is "wrong" for the Foundation Center to charge for access to the information they have collected. It must take a small army of researchers to keep up with all of the information that is available. But for nonprofits without the means to pay for their own subscription or who are unable to access it through a third party, we must find another means to find information on appropriate foundations.

USING ONLINE SEARCH ENGINES

While the trouble with using the Foundation Center's information is cost, the trouble with searching on your own is that, according to the Foundation Center, only 10 to 15 percent of foundations have websites. Thus, you are not necessarily going to get as many results using this approach as you may think you're going to. You are also not going to gain access to as much information as the FDO can give you if you use a search engine approach.

There are three guidelines when using a search engine such as www.Google.com to find appropriate foundations. First, you must be willing to take a lot of time to search. Second, you must try many, many search terms to ensure that you are getting all of the possible foundations. Third, you should use more than one search engine. While Google is the most used search engine in the United States, there are many others to choose, such as Bing (www.bing.com) and Ask (www.ask.com), that may give you slightly different results.

To show the iterative nature of this process, I have done a search using Google with the search term "*foundation violence against women*." (Note that your search will almost certainly not produce the same as shown here, or even the results that someone else reading this will find. Search results change hour by hour, and even depend to some extent on what else you've recently been looking for in the search engine.) Google returned about 5.2 million results (see Figure 3.1). This doesn't mean there are that many foundations with violence against women as a key part of their funding, but the results indicate that there may be quite a few web pages that are connected to this search term in a meaningful way. For comparison purposes, when the search term I used was "violence against women foundation" there were 7.3 million results, fully two million more results, and the first results were not the same as in

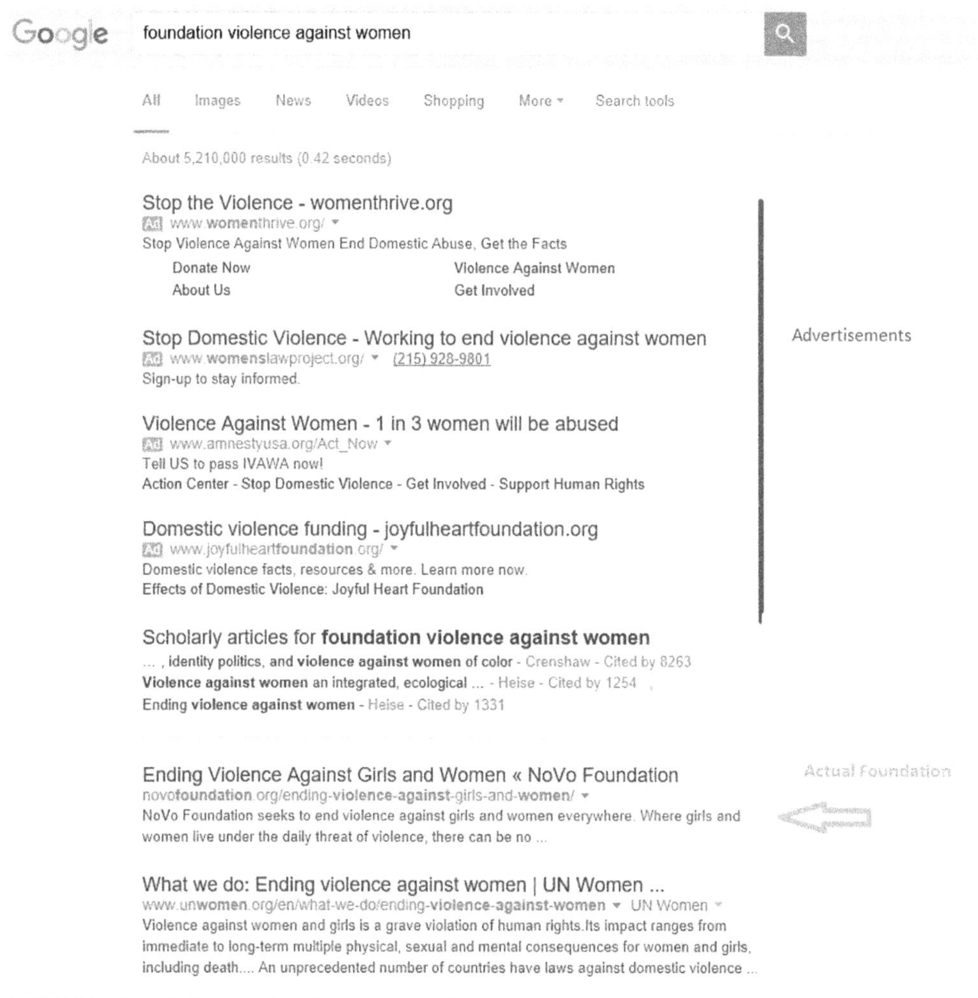

FIGURE 3.1: Search Results for "Foundation Violence Against Women" in Google

the previous search. In a third search, I entered foundation + "violence against women." This time, Google told me it found nearly 3.3 million results. Again, the first page of results was not the same as the previous searches. You can see why it is important to try several variations of your search terms because just the word order, much less choice of similar words, or use of quotation marks, can have a significant impact on the results you see (see Figure 3.1).

In addition to the number of results, Figure 3.1 shows four advertisements that were triggered by the keyword phrase used. One of them has the word "foundation" in their website address, and it actually does not disburse funds. After that are scholarly articles (probably triggered because I use Google Scholar frequently). After that, there is a listing for an actual grant-giving foundation, called NoVo Foundation (see the arrow).

To continue with this example, I selected the first result from the Google search for the NoVo foundation because I can see that it has an initiative to end violence against girls and women and that's what I am looking for.

Looking around on the NoVo Foundation's website, I found out that this organization was started with funding by Warren Buffet, a notable Wall Street investor, who give his son and his wife the responsibility to wisely use $1 billion of a philanthropic gift. Peter and Jennifer Buffet are the copresidents and copresidents of the board of directors of the NoVo Foundation. The foundation has been active since 2006. The mission of the organization is to "foster a transformation from a culture of domination and exploitation to collaboration and partnership." This sounded like exactly what I was looking for. I clicked further into the website only to find some unhappy news for any nonprofit looking for new funding in this area:

> NoVo has dedicated itself to a few very targeted initiatives focused on empowering adolescent girls, ending violence against girls and women, advancing social and emotional learning, and promoting local living economies. Over the next few years, we are working with a strategic set of partners around these initiatives and therefore are not accepting additional requests for funding. (Novo Foundation, 2016)

(Notice here how the NoVo Foundation is being very targeted, reaching out to just a few nonprofits to fund and not accepting unsolicited applications, the three reactions to the Age of Scarcity mentioned earlier in the chapter.)

While it is disappointing that the NoVo Foundation is not accepting unsolicited applications at this point (and they are not soliciting any additional grantees), there are a number of other foundations available to look at from the Google search I did. The next one I chose is the Avon Foundation for Women, which was another result in the search. Their website is at https://www.avonfoundation.org/. They are strong supporters of ending violence against women around the globe. Avon Foundation initiated a global grants program called Speak Out Against Domestic Violence in 2004, and it has donated nearly $60 million since then (https://www.avonfoundation.org/grants/domestic-violence-grants/).

The Avon Foundation presents short information on grant winners in their grant archives, which is helpful to get a sense of the type of programs funded. Curiously, it is not possible to find out when the next round of applications is due as I write this in early June. The website states that they are working with current grantees and to check the website for future opportunities. Based on information from prior years, the application period for this grant has been June 20 to August 1 of each year. This is a short amount of time, so only nonprofits that were already aware of the grant opportunity and when it occurred ahead of the announcement would be likely to submit an application in a timely way. This example shows the power of keeping track of foundation grant opportunities over time, as it allows for better predictions of future opportunities. While such guesses are not always going to be correct, the past provides clues to the future. But I drew another blank in terms of a current request for proposals.

I went to the search results again and found the Mary Kay Foundation. It has a shelter grant opportunity that gives away over $3,000,000 each year, to at least one shelter in every state (Mary Kay Foundation, 2016). The site indicates that the application period is from January 15 to April 30 of each year, with awards being announced in October. A savvy grant-writer interested in a grant for this purpose would write this information down in a file to be reminded of at the start of a year.

CONCLUSION

You should take away these five key lessons from this chapter regarding finding foundation funding. First, the Foundation Center's online directory is most likely your best source for information on foundation grants. They have collected vast amounts of information that can be beneficial for your efforts. Because it is too costly for many nonprofits to order on their own, grantwriters should try to find a ***free*** way to use it, through a public library or other source, such as a nonprofit resource center.

Second, whether using the Online Directory or your own online searches, constantly search for foundations that are active in areas you would like to apply for. Remember, applications may only be allowed for a short period once each year, so you must keep note of when recurring deadlines are. Set up a "tickler" file or notices that will remind you to look up current proposal deadlines and requirements.

Third, knowing that foundations are more and more interested in working with a smaller number of nonprofits they already are aware of, you'll need to be more proactive than ever before. Foundations can, and do, alter their own rules about allocating their funds. Be sure to publicize your organization's efforts year-round, targeting the people who make foundation funding decisions for your information and outreach efforts whenever possible.

Fourth, your organization should be conducting and publicizing rigorous program evaluation results, overseen by outside evaluators. Foundations are looking for successful nonprofits to partner with and soliciting applications only from them. You need to have evidence of effectiveness when you contact (or are contacted by) potential funders.

Fifth, keep an easily referenced set of notes on as many foundations as you can keep track of. Use for a template the questions and pages on the next few pages and regularly search for additional foundations to add to your list. Subscribe to any RSS or e-mail lists from those foundations to keep up to date on what they are doing and whom they are awarding grants to. At some point your organization's name will be listed!

PRACTICE WHAT YOU'VE LEARNED

The application exercises for this chapter are very straightforward but extremely important if you wish to apply for funding from foundations.

STEP 1

Research where you can use the Foundation Center's online directory for free. Are there libraries or nonprofit consortia nearby where you can gain access? List where these are below. Add a telephone number or e-mail address of the person or department where you can make contact to be trained and use the database.

Foundation Center Online Directory Access Point 1: ______________________________

Location: ______________________________

Person or Department Name of Contact: ______________________________

Telephone, e-mail, or URL: ______________________________

Foundation Center Online Directory Access Point 2: ______________________________

Location: ______________________________

Person or Department Name of Contact: ______________________________

Telephone, e-mail, or URL: ______________________________

Foundation Center Online Directory Access Point 3: ______________________________

Location: ______________________________

Person or Department Name of Contact: ______________________________

Telephone, e-mail, or URL: ______________________________

STEP 2

Select one set of keywords to search on that is relevant to your organization. Find at least five foundations that you might be able to request funding from. Locate and write down the answers to the below questions for each foundation and grant program. Repeat with other relevant keyword terms. The more you do this, the better your personal database for foundation funding will be.

Key word search term #1:

____

Name of Foundation #1:

____

URL of Foundation #1:

____

Name of Grant Program #1:

____

URL of Grant Program #1:

____

Dates of Application #1:

____

Other Important Information for #1:

____

____

____

____

____

____

____

____

____

Key word search term #2:

____

Name of Foundation #2:

____

URL of Foundation #2:

____

Name of Grant Program #2:

____

URL of Grant Program #2:

____

Dates of Application #2:

____

Other Important Information for #2:

____

____

____

____

____

____

____

____

____

____

____

____

Key word search term #3:

Name of Foundation #3:

URL of Foundation #3:

Name of Grant Program #3:

URL of Grant Program #3:

Dates of Application #3:

Other Important Information for #3:

REFERENCES

Denver Public Library (2013, May 29). Foundation Center Directory Tutorial. Retrieved from http://youtu.be/tKlb8iurAK4

Foundation Center (2016). We Can Give You Results Google Can't. Retrieved from https://subscribe.foundationcenter.org/fdo

Foundation Center (2014, June 9). Have Foundations Recovered from the Great Recession? Retrieved from http://pndblog.typepad.com/pndblog/2014/06/have-foundations-recovered-from-the-great-recession.html

Gates Foundation (2014). Who We Are: Foundation Fact Sheet. Retrieved from http://www.gatesfoundation.org/Who-We-Are/General-Information/Foundation-Factsheet

Henry J. Kaiser Family Foundation (2014). About Us. Retrieved from http://kff.org/about-us/

Mary Kay Foundation (2016). Shelter Grant Program. Retrieved from http://www.marykayfoundation.org/Pages/ShelterGrantProgram.aspx

Minnesota Council on Foundations (MCF) (2014). What's a Foundation or Grantmaker? Retrieved from http://www.mcf.org/nonprofits/what-is-a-foundation

Novo Foundation (2016). Frequently Asked Questions. Can Our Organization Receive Funding from Novo Foundation? Retrieved from http://novofoundation.org/about-us/faqs/page/2/

Techsoup (2012, September 10). Webinar: Finding Funding with the Foundation Center—2012-09-06. Retrieved from http://youtu.be/PzFu13s-6CQ

CHAPTER 4

FINDING GOVERNMENT FUNDING SOURCES

This chapter discusses the primary ways of finding government funding sources, particularly at the federal level. While national government grants are perhaps the easiest to research, they are also therefore much more competitive to receive. Nonprofits looking for government funding need to be able to research and find ways to locate state and local government grant opportunities as well. The principles described in this chapter work at the state and local levels, although the details vary from location to location.

Before we discuss the specifics of the federal government's available grants, let's look briefly at the impact of the Age of Scarcity on government funding, which was introduced in chapter 2 and also examine the differences between a grant and a contract.

HOW HAVE GOVERNMENTS BEEN AFFECTED BY THE AGE OF SCARCITY?

In many ways, government spending decisions at all levels define the Age of Scarcity, from the automatic budget cuts known as "sequestration" that went into effect on January 3, 2013, to the many years of funding difficulties at the state and local levels since the economic recession hit. The Bureau of Labor Statistics (2012) indicates the duration of the Great Recession was from December 2007 to June 2009. It notes that many statistics have "yet to return to their pre-recession values" (p. 1). Even today, years after the end of the economic downturn, some states are having financial difficulties, which affects their ability to fund nonprofits providing human services.

Federal government spending has been reduced in some areas and is slated to be cut in the future as well. In addition, there is considerable political pressure to maintain or decrease tax rates on Americans and thus keep government spending in check. While stimulus spending at the national level kept government expenditures growing during the recession, this source of spending has ended.

According to the General Accounting Office (2014, pp. 15–16), the effects of sequestration's budget stringency in 2013 included:

- A decrease of about 142,000 participants in the Women, Infants and Children Farmers' Market Nutrition Programs;
- A reduction of 42,000 in the number of very-low-income households that participated in the Housing Choice Voucher program;
- The removal of 60,000 formerly homeless persons from housing and emergency shelter programs; and
- About 57,000 fewer children in Head Start.

States continue to keep a tight lid on their expenditures in order to keep from raising taxes. According to the National Association of State Budget Officers (2015), 42 states expected to spend more in fiscal year 2015 compared with fiscal year 2014, but "many states will continue to face difficult budgetary choices in fiscal 2015 and beyond, since revenue growth may not be sufficient to cover increased spending in many areas" (p. vii). Some states, such as Kansas and Illinois, have cut taxes so steeply that they are essentially starving government of funds to promote basic services. Budget reductions in Kansas in 2015 included cuts to, or no adjustments for inflation for, programs such as Head Start, and in-home services to low-income frail elders at risk of moving to nursing homes. Community development disability organizations saw reduced funds. These organizations assist families to find resources to keep people with disabilities in homes rather than in state institutions. Foster care providers received no cost of living increases (Ranney, 2015). Nonprofits in Illinois have been victimized by the lack of a state budget to pay contract obligations for over a year (Mueller, 2016).

Local governments find themselves in a position that is more constrained than states do. Many states have reduced aid to localities, and local governments face difficulties maintaining infrastructure and basic services (National Association of State Budget Officers, 2015). Over half of states have K-12 schools with less money than in 2008; some states continue to reduce state aid to local school districts (Center on Budget and Policy Priorities, 2016). Local governments rely heavily on property taxes. While housing prices have rebounded in recent years, it takes a long time for all properties to be reassessed and their higher values taxed.

Despite the issues relating to government funding of human services through grants and contracts, nonprofits continue to need this source of income in order to carry on their work. According to the Nonprofit Finance Fund, only 22 percent of the 1,600 human service nonprofits surveyed had plans to replace federal and state government funding with nongovernment income (Nonprofit Finance Fund, 2015). This is true even though about one-fourth of the respondents received less funding from government sources compared to three years earlier (Nonprofit Finance Fund). Grantwriters who operate in this brutal environment will likely experience greater stress, lack of job satisfaction, burnout, and feelings of personal failure if their proposals are not funded, even though the environment is making success more difficult. *Grantwriters need to become more skilled or else risk the survival of their organizations.* If other grantwriters are improving and they are not, they are actually falling behind in their skill sets.

WHAT IS THE DIFFERENCE BETWEEN A GOVERNMENT GRANT AND A CONTRACT?

Nonprofit leaders may talk about receiving a government grant or a government contract as if they are the same thing. While both bring funding to an organization, they are different in many important ways. This section looks at the main ways they differ (see Table 4.1).

First, the purposes and goals of a grant and a contract are not the same. A grant has the purpose of transferring "money, property, services or anything of value to a recipient" (31 U.S.C. section 6304), whereas the purpose of a contract is to "obtain (purchase, lease or barter) goods (property) or services" (31 U.S.C. section 6303). A grant's goal is to "accomplish a public purpose," while the goal of a contract is to achieve something "for the direct benefit of the U.S. Government."

Once a grant is awarded, there is only minor involvement with the operations of the grantee, and the grantee is given fairly free rein to accomplish the public purpose. Grantees must provide required reports and respond to performance issues. These are the main ways the government maintains contact with the grantee. When a contract is awarded, government still has major involvement in the terms of the contract and how it achieves its purposes. Definite tasks, milestones, and deliverables expected are part of the contract between government and agency.

Nonprofits generally set their scope of work in responding to the government agency's request for proposals (RFP). While the RFP lays out the general parameters of the applications, each responding nonprofit may define their response as they want. With a contract,

TABLE 4.1: Differences Between Federal Government Grants and Contracts

Characteristic	Grant	Contract
Purpose	Transferring money, property, services, or anything of value to a recipient (31 U.S.C. Section 6304)	Obtaining (purchase, lease, or barter) goods or services (31 U.S.C. Section 6303)
Goal	Accomplishing a public purpose	Achieving something for the direct benefit of the US Government
Amount of Government Involvement Once Awarded	Minor	Major
Who Sets Scope of Work?	Nonprofit	Government
How Regulated?	Through OMB Circulars A-21 (2 CFR 220) and A-110 (2 CFR 215)	Through Federal Acquisition Regulations (FAR) and Federal Agencies FAR Supplements

Source: Adapted from Doreen Woodward (2011). Grants and Contracts: How They Differ. Michigan State University. Retrieved from https://www.cga.msu.edu/PL/Portal/DocumentViewer.aspx?cga=aQBkAD0AMgAxADQA

on the other hand, the government is explicit in what it wants and nonprofits have limited scope in determining what to offer other than the stated goods or services. Finally, grants are regulated through OMB Circulars A-21 (2 CFR 220) and A-110 (2 CFR 215), whereas contracts are regulated by the Federal Acquisition Regulations (FAR) and Federal Agencies FAR Supplements.

Examples of the two funding mechanisms can help clarify the situation. A grant mechanism was used by the federal government agency, the Administration for Children and Families, Office of Community Services (OCS), to "address food deserts; improve access to healthy, affordable foods; and address the economic needs of low-income individuals and families through the creation of business and employment opportunities" (HHS-2013-ACF-OCS-EE-0584). This does not have a fixed price, but agencies can decide how much to request, up to the maximum amount allowed ($800,000) for up to five years. More specifically, the OCS seeks to fund projects that will implement innovative strategies for increasing healthy food access while achieving sustainable employment and business opportunities for recipients of Temporary Assistance for Needy Families (TANF) and other low-income individuals whose income level does not exceed 125% of the federal poverty level.

When talking about a grant, the government agency states the general purpose, but the way to achieve the stated purpose is left up to the organization making application. Thus, a considerable amount of creativity is required within the grant application. This is a grant.

An example of a contract is the Centers for Medicare and Medicaid Services signing an agreement for "Administrative Management and General Management Consulting Services" with the Health Research and Educational Trust for almost $76 million a few years ago.

The rest of this chapter examines how to use the primary research tool to find grants (not contracts) at the federal government level, www.grants.gov . If you live in a state that has a similar system the information will be applicable there as well (an example is Texas's Texas eGrants Search, https://www.texasonline.state.tx.us/tolapp/egrants/search.htm).

The fact that all federal government grants can be accessed through one portal is a great boon to grant seekers. It does, however, place a premium on your ability to conduct an effective search there.

USING THE FEDERAL GOVERNMENT'S GRANTS.GOV WEBSITE

If you have access to a computer while you're reading this, you will find it valuable to check out some of the training materials available online. I've developed a quick step-by-step video that's posted on YouTube called "How to Use Grants.gov to Find Federal Grants," which is available at http://youtu.be/yDbGerr5Oek. Other videos with helpful information are available on the www.grants.gov website and on YouTube if you search "How to use grants.gov."

This chapter contains screenshots of various pages to help you learn how to navigate through the pages. Because the grants.gov website changes from time to time, what you see when you log on may not be exactly the same as what is shown here. Still, the basic features

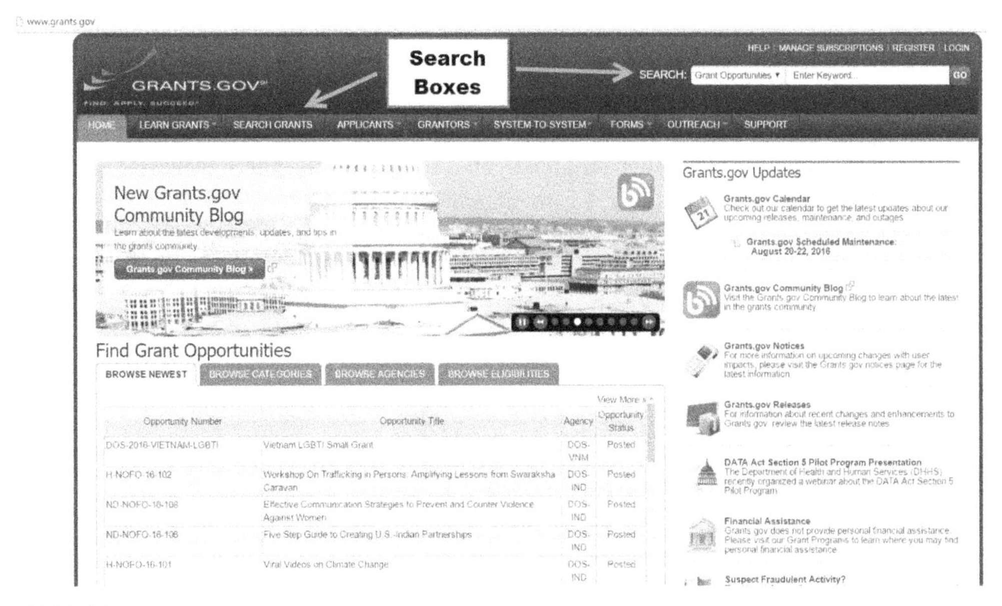

FIGURE 4.1: Grants.gov Home Page

have remained very similar over the past few years, so it is likely you can find what is being described here. Because of the nature of what grant opportunities are available at any one time, there may not be any requests for proposals that interest you when you look. If this is true, choose another topic so you can learn the process to help you build your expertise in using the grants.gov database.

Figure 4.1 shows the home page. The arrows drawn near the top point to the tab that says "search grants"—this is where you want to click. Find it and click once. You'll see a new screen, as in Figure 4.2. (You can also search directly by using the search box at the upper right of the screen.)

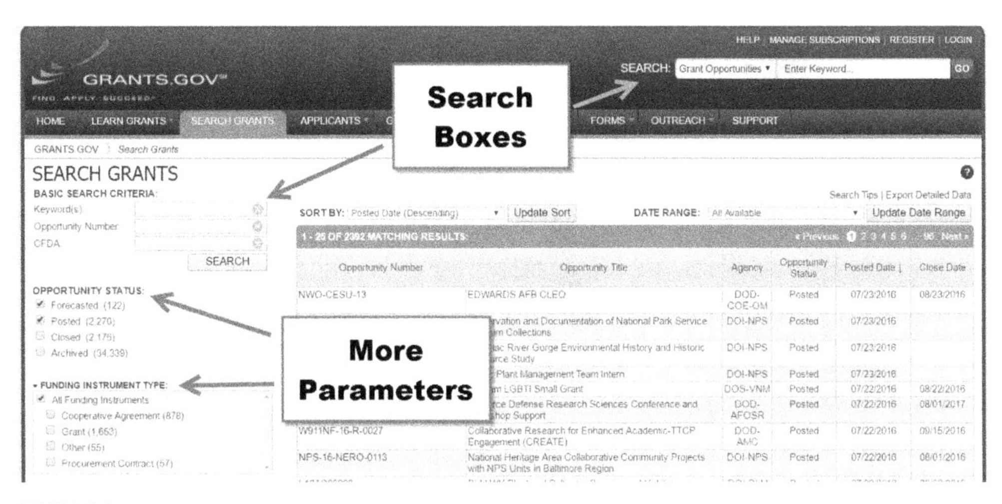

FIGURE 4.2: Grants.gov Search Grants Page

As shown in Figure 4.2, you have two places you can enter your keywords for the subject of your grant search. The first is on the left side of the screen, and the other is at the top on the right (see arrows in Figure 4.2). Type your search term in either location. But don't hit ENTER yet—there is more to do before beginning the actual search. We want to set additional parameters to achieve more targeted results. These parameters run along the left-hand side of the grants.gov page, as also shown in Figure 4.2).

For this example, we use the keywords "violence against women" in either of the search boxes. You may also search by Funding Opportunity Number or by CFDA (Catalog of Federal Domestic Assistance) number, if you know what these are. Below the "CFDA Number" box are more potential search parameters you can select (see Figure 4.2 and Figure 4.3).

In Figure 4.3, the keywords "violence against women" are typed in (see arrow 1). When you have typed in the keyword, look below what you have typed (see arrow 2). The boxes there tell the database how far back in time to search for funding opportunity announcements (FOAs). The topmost box ("Forecasted") provides advance information about grant opportunities that agencies *believe* will be open in the future. This is a new feature as of June 2016. At the time this screenshot was taken, there were zero forecasted grants on this topic. Checking the second box ("Open") shows only FOAs that are currently available and accepting applications. This is the box that most grantwriters are used to using. In this example, grants.gov finds 42 open grants using the search term "violence against women."

Checking the third box adds the FOAs that are no longer accepting applications and have been closed 30 or fewer days. This yields 21 results. The final box ("Archived") shows FOAs that have been closed for more than 30 days—399 results are indicated at this time. You may be wondering why anyone would want to see results that include closed and archived FOAs. This is discussed later, as it is an important way of learning about grant opportunities that may come up regularly.

Continuing to look at Figure 4.3, find arrow 3. It points to the type of funding instrument you are searching for. You may choose to include cooperative agreements or not. I have chosen to check only "Grants" for this example. Arrow 4 points to the type of organizations that are eligible. Here I have checked the box that indicates the applicant may be a 501c3 (charitable nonprofit) registered with the IRS. There are many options here, so you will want to look at the list and check all that apply to you.

Also in Figure 4.3, we have options concerning the funding category (see arrow 5). This relates to the category of funding—that is, the general area of government activity. For this example, "violence against women," the most logical category is "Law, Justice and Legal Services," so that has been chosen. Two other categories are also checked: "Health" and "Income Security and Social Services." If you are unsure, look at the list carefully. It is perhaps better when you are starting out that you choose too many categories rather than too few. The number in the parentheses indicates how many grant opportunities are active with that option checked.

The final area of search options is "Agency" (arrow 6). This refers to the precise government agency administering the grant. Many federal government agencies are listed, so it is important to try to find the one(s) most likely to offer grants in the area you are searching. This is not always obvious, so it is often useful to go too wide than too narrow when you are starting out. "Violence against women" has several different options that are

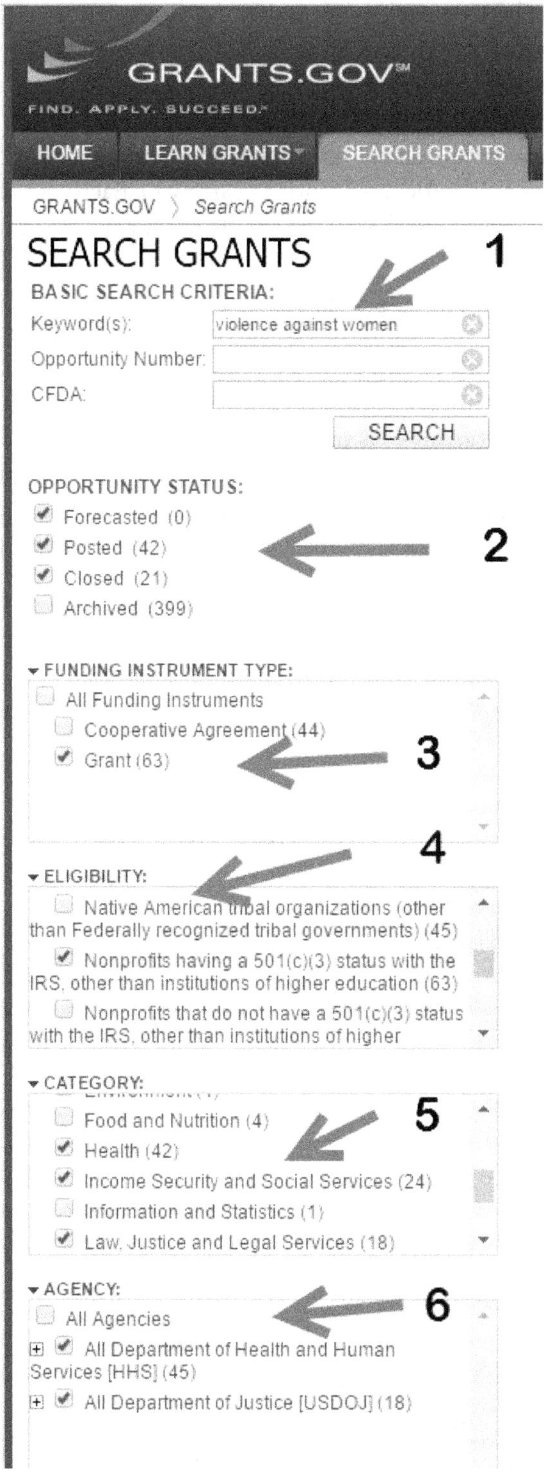

GRANTS.GOV℠
FIND. APPLY. SUCCEED.℠

HOME | LEARN GRANTS | SEARCH GRANTS

GRANTS.GOV 〉 *Search Grants*

SEARCH GRANTS

1

BASIC SEARCH CRITERIA:

Keyword(s): violence against women

Opportunity Number:

CFDA:

SEARCH

OPPORTUNITY STATUS:

- [x] Forecasted (0)
- [x] Posted (42)
- [x] Closed (21)
- [ ] Archived (399)

2

FUNDING INSTRUMENT TYPE:

- [ ] All Funding Instruments
  - [ ] Cooperative Agreement (44)
  - [x] Grant (63)

3

4

ELIGIBILITY:

- [ ] Native American tribal organizations (other than Federally recognized tribal governments) (45)
- [x] Nonprofits having a 501(c)(3) status with the IRS, other than institutions of higher education (63)
- [ ] Nonprofits that do not have a 501(c)(3) status with the IRS, other than institutions of higher

CATEGORY:

- [ ] Food and Nutrition (4)
- [x] Health (42)
- [x] Income Security and Social Services (24)
- [ ] Information and Statistics (1)
- [x] Law, Justice and Legal Services (18)

5

AGENCY:

- [ ] All Agencies
- [x] All Department of Health and Human Services [HHS] (45)
- [x] All Department of Justice [USDOJ] (18)

6

FIGURE 4.3: Search Options

strong possibilities. The Department of Health and Human Services and the Department of Justice are two agencies that seem most likely to have open FOAs. Some agencies have "subagency" options that you may check or uncheck. This is not shown in Figure 4.3, but is sometimes helpful to more quickly narrow your search.

Figure 4.4 shows what the actual results are, after adding in "Archived" FOAs. First, a total of 462 funding opportunities are identified (arrow 1). Each page shows 25 results, so be sure to check that you have looked at all the results when you scroll through. (Note: Figure 4.4 shows only 17 of the 25 results on the page, in order to be enlarged.)

Arrow 2 points to the heading "Opportunity Status," letting you know whether the grant window will start in the future (forecasted), is open now (open), was closed recently (closed) or has been closed for a long time (archived). Only the options checked have results visible. "Posted Date" (arrow 3) shows when the funding opportunity was announced by the agency. "Close Date" (arrow 4) is the date the opportunity closed. Because this screenshot was taken in late July 2016, the FOAs shown here were posted in March, April, May, June, and July. Closing dates may be as little as four to six weeks after posting, or up to several years, as shown by the variety of time lengths here. This page of results shows that there

Search Tips | Export Detailed Data

SORT BY: Posted Date (Descending) Update Sort DATE RANGE: All Available Update Date Range

1 - 25 OF 462 MATCHING RESULTS: ← 1   2 →   3   4   « Previous 1 2 3 4 5 6 ... 19 Next »

Opportunity Number	Opportunity Title	Agency	Opportunity Status	Posted Date	Close Date
RFA-OD-16-013	Building Interdisciplinary Research Careers in Women's Health (K12)	HHS-NIH11	Posted	07/14/2016	10/06/2016
PA-16-313	Safety and Outcome Measures of Pain Medications Used in Children and Pregnant Women (R03)	HHS-NIH11	Posted	06/07/2016	01/07/2020
PA-16-312	Safety and Outcome Measures of Pain Medications Used in Children and Pregnant Women (R21)	HHS-NIH11	Posted	06/07/2016	01/07/2020
PA-16-311	Safety and Outcome Measures of Pain Medications Used in Children and Pregnant Women (R01)	HHS-NIH11	Posted	06/07/2016	01/07/2020
HHS-2016-ACF-ACYF-SR-1197	Sexual Risk Avoidance Education Program	HHS-ACF-FYSB	Posted	05/26/2016	07/25/2016
PA-16-289	Research Supplements to Promote Re-Entry into Biomedical and Behavioral Research Careers (Admin Supp)	HHS-NIH11	Posted	05/24/2016	09/29/2019
RFA-AI-16-047	Partnerships for Structure-Based Design of Novel Immunogens for Vaccine Development (R01)	HHS-NIH11	Posted	05/18/2016	10/03/2016
OJJDP-2016-10040	OJJDP FY 2016 Studies Program on Trauma and Justice-Involved Youth	USDOJ-OJP-OJJDP	Closed	05/09/2016	06/23/2016
HHS-2016-ACF-ACYF-CY-1121	Basic Center Program	HHS-ACF	Closed	05/05/2016	07/05/2016
PA-16-188	Mechanisms, Models, Measurement, and Management in Pain Research (R01)	HHS-NIH11	Posted	04/14/2016	05/07/2019
PA-16-187	Mechanisms, Models, Measurement, and Management in Pain Research (R21)	HHS-NIH11	Posted	04/14/2016	04/14/2019
HHS-2016-ACF-ACYF-CA-1119	Grants to Tribes, Tribal Organizations, and Migrant Programs for Community-Based Child Abuse Prevention Programs	HHS-ACF	Closed	04/06/2016	06/27/2016
NIJ-2016-9600	Research and Evaluation on Victims of Crime	USDOJ-OJP-NIJ	Closed	03/30/2016	06/03/2016
HHS-2016-ACL-NIDILRR-DP-0145	Disability and Rehabilitation Research Projects (DRRP) Program: Traumatic Brain Injury (TBI) Model Systems National Data and Statistical Center	HHS-ACL	Archived	03/29/2016	05/31/2016
RFA-AI-16-034	Partnerships for Countermeasures against Select Pathogens (R01)	HHS-NIH11	Posted	03/29/2016	10/03/2016
OVW-2016-9780	OVW FY 2016 Research and Evaluation Program	USDOJ-OJP-OVW	Archived	03/21/2016	05/02/2016
NIJ-2016-9235	Research on "Sentinel Events" and Criminal Justice System Errors	USDOJ-OJP-NIJ	Closed	03/18/2016	05/17/2016

FIGURE 4.4: Grants.gov Search Results Page, Violence Against Women Search Term

are no open grant opportunities that are of interest to an agency looking for funding to run a program, although a number of grants may be of interest for health-focused researchers. These have the designations R01, R03, and R21 after the opportunity title.

WHAT TO DO WHEN THERE ARE NO GRANTS TO APPLY FOR?

At first glance, the results found in Figure 4.4 are discouraging. No grant opportunities fitting what we want to write about are open, and none are forecasted to open up in the next few months. This doesn't mean, however, that we are done with grants.gov. We can learn valuable information from what is hidden in the "closed" and "archived" categories of grant opportunities. To illustrate this, Figure 4.5 shows what occurs when we drill down into the information further.

1 - 25 OF 37 MATCHING RESULTS: « Previous 1 2 Next »

Opportunity Number	Opportunity Title	Agency	Opportunity Status	Posted Date ↓	Close Date
OVW-2016-9780	OVW FY 2016 Research and Evaluation Program	USDOJ-OJP-OVW	Archived	03/21/2016	05/02/2016
OVW-2016-9206	OVW FY 2016 Improving Criminal Justice Responses to Sexual Assault, Domestic Violence, Dating Violence, and Stalking Grant Program (also known as the Arrest Program)	USDOJ-OJP-OVW	Archived	01/07/2016	03/03/2016
OVW-2016-9149	OVW FY 2016 Grants to Enhance Culturally Specific Services for Victims of Sexual Assault, Domestic Violence, Dating Violence and Stalking Program Solicitation	USDOJ-OJP-OVW	Archived	12/21/2015	02/24/2016
OVW-2016-9148	OVW FY 2016 Legal Assistance for Victims Grant Program	USDOJ-OJP-OVW	Archived	12/15/2015	02/08/2016
OVW-2016-9104	OVW FY 2016 Rural Sexual Assault, Domestic Violence, Dating Violence and Stalking Program	USDOJ-OJP-OVW	Archived	12/08/2015	02/01/2016
OVW-2016-9003	OVW FY 2016 Justice for Families Program	USDOJ-OJP-OVW	Archived	11/24/2015	01/21/2016
OVW-2015-4031	OVW FY 2015 Grants to Encourage Arrest Policies and Enforcement of Protection Orders Program	USDOJ-OJP-OVW	Archived	02/11/2015	03/24/2015
OVW-2015-4037	OVW FY 2015 Rural Sexual Assault, Domestic Violence, Dating Violence and Stalking Program	USDOJ-OJP-OVW	Archived	01/21/2015	03/04/2015
OVW-2015-4036	OVW FY 2015 Legal Assistance for Victims Grant Program	USDOJ-OJP-OVW	Archived	01/14/2015	03/11/2015
OVW-2015-4033	OVW FY 2015 Grants to Enhance Culturally Specific Services for Victims of Sexual Assault, Domestic Violence, Dating Violence and Stalking Program	USDOJ-OJP-OVW	Archived	01/06/2015	02/19/2015
OVW-2015-4035	OVW FY 2015 Justice for Families Program	USDOJ-OJP-OVW	Archived	12/31/2014	02/11/2015
OVW-2014-3690	OVW FY 2014 Legal Assistance for Victims Grant Program-Sexual Assault and/or Tribal Focus	USDOJ-OJP-OVW	Archived	12/13/2013	01/30/2014
OVW-2014-3695	OVW FY 2014 Rural Sexual Assault, Domestic Violence, Dating Violence and Stalking Assistance Program	USDOJ-OJP-OVW	Archived	12/13/2013	01/23/2014
OVW-2014-3699	OVW FY 2014 Grants to Enhance Culturally Specific Services for Victims of Sexual Assault, Domestic Violence, Dating Violence and Stalking Program Solicitation	USDOJ-OJP-OVW	Archived	12/12/2013	02/05/2014
OVW-2013-3411	OVW FY 2013 Grants to State and Territorial Sexual Assault and Domestic Violence Coalitions and Sexual Assault Service Programs for State and Territorial Coalitions	USDOJ-OJP-OVW	Archived	04/09/2013	05/22/2013
OVW-2013-3408	OVW FY 2013 Tribal Domestic Violence and Sexual Assault Coalitions Program	USDOJ-OJP-OVW	Archived	03/14/2013	04/25/2013
OVW-2013-3396	OVW FY 2013 Rural Sexual Assault, Domestic Violence, Dating Violence and Stalking Assistance Program	USDOJ-OJP-OVW	Archived	02/07/2013	03/26/2013
OVW-2013-3404	OVW FY 2013 Legal Assistance for Victims Grant Program	USDOJ-OJP-OVW	Archived	02/05/2013	03/19/2013

FIGURE 4.5: Results from Office of Violence Against Women

The list in Figure 4.5 shows open dates covering several years' worth of archived grants, from 2013 to 2016. The bulk of the grant opportunities are posted from late November through mid-March, no matter which year one examines. They tend to close in February, March, and April. One reason for the similar opening dates is that these eight listings really are many of the same funding opportunities but for different years. This brings up the concept of a "grant season," which means that grant opportunities in a particular field of practice may be concentrated in a few months of the year. Looking at a list such as Figure 4.5 demonstrates that for "violence against women" grants, the season when grants are due is in the spring of the year. Understanding when the grant season is in the field for your organization is important, so you can plan properly.

Thus, as a grantwriter, even if you are too late to apply in any one year, you can use the information in your www.grants.gov search to lay out a calendar for checking back each year. You also should note if two separate funding programs have their open and due dates very close to each other. If you wanted to write a grant proposal for both of these opportunities, you will have a lot of work to do to write two major grants in a few weeks.

Looking at Figure 4.5, we see additional information of value. For example, suppose you want to write a grant for a dating violence agency in a rural part of your state. A grant program exists specifically for these programs, called the "Rural Sexual Assault, Domestic Violence, Dating Violence and Stalking Assistance Program." The arrows in Figure 4.5 point to the listings for this program for several years, from 2013 to 2016. This tells grantwriters that the program has been in existence for at least several years and so one might believe it will be in place for 2017 and onward as well.

In planning to write a proposal for this grant, you could notice when the opportunity has been posted and when it has been due across these years. For example, in 2013, the proposal was open on February 7 and closed on March 26. For the 2014 competition, posting occurred earlier, already on December 13, 2013, closing on January 23, 2014. In 2015, posting was done on January 21. The opportunity closed on March 4. Finally, we see the grant was posted on December 8, 2015, for the 2016 funding year. Closing happened on February 1, 2016. While these dates all fall into the "grant season" noted earlier, there is a bit of variation that makes planning for the next year an "approximate thing."

You can look at the grant opportunities in other rows and notice that some of them have only been issued for one or two years. Some are seemingly no longer offered, as they haven't been posted for a couple of years. This shows that the federal government changes grant opportunities, bringing in new ideas and scaling back or eliminating older ideas. Particularly when a new person assumes office as president or even just within an agency, we see shifts in priorities regarding use of discretionary funding. To the extent that the past predicts the future, however, it is helpful to examine closed and archived grant opportunities.

Sometimes a shift in priorities leads to a change in the name of the funding opportunity. In 2014, a grant program called "Legal Assistance for Victims Grant Program—Sexual Assault and/or Tribal Focus" was launched. In 2015, in what might be a similar program, there was the "Legal Assistance for Victims Grant Program," without any particular focus. This may indicate a small change in priorities, or it may not. Only reading the details of the two funding opportunities could answer the question for sure.

WHAT ELSE CAN WE LEARN?

We can go one step deeper in using the information from grants.gov, even if there are no active grant opportunities. When looking at search results in grants.gov, you can click on the blue underlined funding opportunity title, which is a hyperlink. You are then taken to a page with details about the grant opportunity. Among other information you can immediately access are the award ceiling (how much a grant proposal can be for), the expected number of awards (to judge your chances a bit, assuming all else is equal), and the total amount for all awards. You can compare this information for the same grant opportunity in different years to see what trends there may be in the funding stream.

Using the "Rural Sexual Assault, Domestic Violence, Dating Violence and Stalking Program" grant opportunity from the Department of Justice, Office on Violence Against Women, we can compare five years of information, with the requests from proposals from fiscal year (FY) 2012 through FY 2016 (United States Department of Justice, 2012, 2013, 2014, 2015, 2016). (See the listing of hyperlinks for each of the Request for Proposals for the various years in the "References" section of this chapter, noting that hyperlinks may change over time).

In FY 2012, the competition was limited to organizations that were already being funded from the FY 2009 or FY 2010 competition years. Recipients from FY 2011 were not eligible. No other organizations were allowed to apply. The number of anticipated continuation grants was not specified, nor was amount of total funding noted. Awards of more than $1,000,000 for a three-year period were "unlikely."

For the FY 2013 awards, the competition was opened up to include "new applicants who have never received funding under the Rural Program or received funding prior to FY 2010" as well as applicants who had received funding in FY 2010. Funding levels were described as in the FY 2012 RFP.

In FY 2014, the competition was available "by invitation only" to grantees who had received funding in FY 2011. By the FY 2015 competition, agencies that had never been awarded funding or that had not received funding in the past 12 months could apply. For the first time, the RFP noted that the Office of Violence Against Women "typically makes awards in the range of $175,000—$1,000,000, and estimated it would make "up to 40 awards for an estimated total of $33,000,000" (p. 12).

The FY 2016 Rural Program cycle was open to both new and continuation applicants. Interestingly, new applications were limited to $500,000 for the three-year award period, while continuation proposals could be for up to $750,000. The typical range of awards had shifted to a range between $350,000 and $750,000. For FY 2016, the number of expected awards grew to "up to 50" with the same total amount of funding, $33,000,000 (p. 7).

The comparison information we uncovered here regarding eligibility, award numbers, and amount of funding per grant is somewhat tedious to track down but is invaluable. If you were interested in this particular grant, you know that new applicants have not been allowed in some years, so if it is an "open year" you should make every effort to apply in case you wouldn't be eligible to apply the next year. You would also note that the Office of Violence Against Women has increased the number of expected grants but decreased the average amount of funding per grantee.

We must remember that these grant opportunities are listed as "discretionary," meaning that they are not awarded by a formula, and also that it is easier for Congress to reduce or eliminate funding for them. As the Age of Scarcity hits harder and harder, it is exactly these types of grant opportunities that are under the most pressure. We see that funding did not drop overall, but individual agencies are being given less to do their work. Total allocations could very well decrease in the upcoming funding cycles, when average grant size might stay the same but with fewer awards made.

Astute grantwriters will take the time to do a careful longitudinal analysis of several years' of the funding opportunity statements. These statements contain all the requirements, goals, and other information you must have in order to write your proposal. A partial table of contents includes:

- Who is eligible to apply;
- Deadlines for submission;
- Whether preapplication conference calls will be held for questions to be asked, and when;
- Contact information;
- Overview of the funding opportunity;
- Award information;
- Program scope, including purpose areas, priority areas, out-of-scope activities, and unallowable activities;
- How to apply;
- Application contents, including formatting and technical requirements, application requirements, project narrative, budget detail worksheet and narrative, and more;
- Additional required information;
- Selection criteria, including review process, past performance review, compliance with financial requirements, and more; and
- Post-award information requirements.

While it sometimes happens that the contents of the funding opportunities change drastically from one year to the next, this is relatively rare. If you look at several years of the same funding announcement and the information is very similar, you have found a way to anticipate what is likely to be requested for the next round of applications. This way, instead of having about six weeks to get your 30- to 50-page application written, you have a year or more, with only small changes to be made based on minor changes in what is being asked for.

Let's compare, for example, the priority areas from FY 2012 to FY 2016 funding opportunity notices for the Rural Grants program that we've examined so far in this chapter (United States Department of Justice, 2012, 2013, 2014, 2015, 2016). While additional priorities come and go, there is one core element that remains nearly identical over the entire time. The similar priorities are shown in Table 4.2.

While there is variation in wording, particularly in FY 2014, the basic concept of one priority of the program for all five years is nearly unchanged. After so many years of keeping a similar wording, it is likely that the Rural Grant program will have at least this one priority area in place for at least a few more years, barring any changes in the underlying legislation.

TABLE 4.2: Priority Statements for OVW Rural Grants Program, FY2012–FY2016

Fiscal Year	Statement
2012	"Projects that focus primarily (80% or more) on a multi-disciplinary effort to improve the criminal justice system's response to sexual assault, including programs that encourage and support the development or enhancement of investigative and prosecutorial efforts" (p. 8).
2013	"Projects that focus primarily (75% or more) on increased support for sexual assault, including services, law enforcement response and prosecution, and those projects proposing to involve the implementation of the Prison Rape Elimination Act of 2012 (PREA) standards in working with incarcerated victims" (p. 11).
2014	"To increase the safety and well-being of women and children in rural areas or rural communities by dealing directly and immediately with sexual assault, domestic violence, dating violence and stalking occurring in rural areas or rural communities; and creating and implementing strategies to increase awareness and prevent sexual assault, domestic violence, dating violence and/or stalking" (p. 8, labeled a "Purpose Area").
2015	"Prioritizing projects that focus primarily (75% or more) on developing, enlarging, or strengthening programs addressing sexual assault, including sexual assault forensic examiner programs, Sexual Assault ResponseTeams, law enforcement training, programs addressing rape kit backlogs, and programs that involve implementation of the Prison Rape Elimination Act of 2012 (PREA) standards in working with incarcerated victims" (p. 9).
2016	"Increase support for sexual assault, including services, law enforcement response and prosecution" (p. 4).

CONCLUSION

This chapter has covered a considerable amount of information. Here are key points to recall and practice.

- Government funding has been particularly hard hit by changing political trends and a shift in vocal public opinion that stresses budget limitations over paying for additional expenditures. Human services, in the Age of Scarcity, are seen as less necessary than before. Despite this, federal grants are still available and can be "game changers" for nonprofits and their clients. It thus remains a priority to learn how to write the best grant proposals possible.
- Differences exist between federal government grants and contracts, with grants providing the applicant with considerable latitude to propose their solutions to the pressing problem to be addressed, as described in the FOA or RFP.
- Grantwriters must learn how to use the advanced search features in grants.gov, the main portal to all federal grants and contracts. Using this process allows you to know which grant opportunities are currently open for applications and which are forecasted to be open in the near future.
- Beyond knowing which opportunities are forecasted and are currently open, using "closed" and "archived" funding opportunities on the grants.gov site allows you to compare several years of grant announcements. Having access to this information

lets you do two important tasks. The first is to improve your planning by putting approximate dates on your calendar for when the next year's announcement will be posted and when the application will be due. If you do this for all the federal grant opportunities you are interested in, you can more effectively make long-range plans. The second task is to compare multiple years of grant application requirements so that you can determine what the likely requirements will be in the next round. Priorities often do not change much from one year to the next. Because one year's change in priority may be the next year's change in requirements, knowing what has stayed the same and what has been altered gives you a fantastic opportunity to get a peek as to what may be coming around the bend before it gets here. If you can be 80–90 perent sure of what the application is going to call for next year, you can already be putting together a coherent grant proposal now, giving yourself far more than six weeks to write a strong grant application.

PRACTICE WHAT YOU'VE LEARNED

1. Go to www.grants.gov on your computer. Follow the steps that are in this chapter and come up with at least three potential grant opportunities. Print off the information or save it electronically.
2. Look at forecasted and current grant opportunities as well as RFPs that have expired and been archived. Create a calendar for yourself with the RFPs that are not now available but have a record of having been put forth for the past year or two.

Keywords used	**Request for Proposal ID number**	**Date released**	**Downloaded and saved information?**

3. Compare several years of RFPs for a grant opportunity you are interested in applying for. What can you learn about posted and closing dates, priorities, and other elements of the grant opportunity? What has changed across the years? What has remained similar? How can this analysis help you in writing a proposal to this funder?

REFERENCES

Bureau of Labor Statistics (2012). The Recession of 2007–2009. Retrieved from http://www.bls.gov/spotlight/2012/recession/pdf/recession_bls_spotlight.pdf

Center on Budget and Policy Priorities (2016, January 25). Most States Have Cut School Funding and Some Continue Cutting. Retrieved from http://www.cbpp.org/sites/default/files/atoms/files/12-10-15sfp.pdf

General Accounting Office (2014, March). 2013 Sequestration: Agencies Reduced Some Services and Investments While Taking Certain Actions to Mitigate Effects. Retrieved from http://www.gao.gov/assets/670/661444.pdf#page=22

Mueller, S. (2016). Rauner Vetoes $3.9B in Higher Ed, Social Services Funding. http://wuis.org/post/rauner-vetoes-39b-higher-ed-social-services-funding#stream/0

National Association of State Budget Officers (2015). The Fiscal Survey of States: An Update of State Fiscal Conditions. Washington, DC: Author. Retrieved from http://www.nasbo.org/sites/default/files/NASBO%20Spring%202015%20Fiscal%20Survey%20of%20States%20-%20S.pdf

Nonprofit Finance Fund (2015). 2015 State of the Nonprofit Sector. Retrieved from http://survey.nonprofitfinancefund.org/?filter=org_type:Human%20Services

Ranney, D. (2015, January 16). Proposed Budget Avoids Cuts in Social Services. Kansas Health Institute. Retrieved from http://www.khi.org/news/article/proposed-budget-avoids-cuts-in-social-services

United States Department of Justice (2012). OVW Fiscal Year 2012 Rural Sexual Assault, Domestic Violence, and Stalking Assistance Program. Retrieved from https://www.justice.gov/sites/default/files/ovw/legacy/2012/02/16/fy2012-rural-solicitation.pdf

United States Department of Justice (2013). OVW Fiscal Year 2013 Rural Sexual Assault, Domestic Violence, and Stalking Assistance Program. Retrieved from https://www.justice.gov/sites/default/files/ovw/legacy/2013/02/26/2013-rural-solicitation.pdf

United States Department of Justice (2014). OVW Fiscal Year 2014 Rural Sexual Assault, Domestic Violence, and Stalking Assistance Program. Retrieved from https://www.justice.gov/sites/default/files/ovw/legacy/2013/12/18/fy-2014-rural-solicitation.pdf

United States Department of Justice (2015). OVW Fiscal Year 2015 Rural Sexual Assault, Domestic Violence, and Stalking Assistance Program. Retrieved from https://www.justice.gov/sites/default/files/ovw/pages/attachments/2015/02/12/ovw_fy_2015_rural_sexual_assault_domestic_violence_dating_violence_and_stalking_program_solicitation_2_12_2015.pdf

United States Department of Justice (2016). OVW Fiscal Year 2016 Rural Sexual Assault, Domestic Violence, and Stalking Assistance Program. Retrieved from https://www.justice.gov/ovw/file/797706/download

CHAPTER 5

UNCOVERING NEED IN YOUR COMMUNITY

Grantwriting presents its practitioners with a "chicken or egg" problem. Which comes first: the grant opportunity or the need for a grant? It doesn't benefit an organization to identify needs if there is no funding to solve them. But it also doesn't benefit an organization to find a lot of grant opportunities if those particular issues are not relevant to its community.

My approach to this issue is reflected in the arrangement of this book: I have presented the chapters on finding grants before this chapter on uncovering needs. I believe there is a great deal of benefit to knowing what is available before you look for needs to deal with or problems to fix. The unfortunate truth is that most communities have a large number of problems. Many areas of our country have high unemployment rates, lack of adequate schooling opportunities, and frequent violence in the neighborhood. Even well-off areas experience substance abuse, teen delinquency and truancy, lack of affordable healthcare, and so on.

Finding problems to fix is not much of a challenge. If you keep your eyes open and talk to constituents and stakeholders, you can uncover needs galore. Grantwriters, however, must be able to link funding opportunities with needs, not just locate one without reference to the other.

The difficult part, then, is to have in mind what issues funders will provide resources to address. It is part of the "think like a funder" approach that I advocate. If you focus on what the problems are in your community that YOU want to deal with, you may find yourself with a great deal of good information on "unfundable" problem topics. If you know which needs are being funded now, you are focusing on the FUNDER'S wants, which will ultimately be more productive in writing a winning grant. Once you know the range of issues and problems that have resources attached, you can begin to determine how much they affect your community.

THE 4 "Ds" OF UNCOVERING NEED

1. **Define the terms** explicitly in the same way as the funder does: What do you mean when you say "Y?"

2. **Discover data** that show the extent of the need in your community.
3. **Describe the need** in your community, possibly compared with other communities.
4. **Detail the impact** of the need or problem on people in your community.

Once you have gone through these four steps, you will have written a compelling statement that uncovers an important need for your community. This is the foundation for writing a successful grant application (Figure 5.1).

STEP 1: DEFINE THE TERMS

One situation that is frequently encountered by grantwriters and program planning personnel is that they "know" there is a problem, but it isn't clear exactly what they mean when they talk about it to others. For example, you might hear a board member say that there is a problem with poverty in your community and that you, as a grantwriter, should work on obtaining resources to combat it. When you question the board member more closely, however, it isn't clear whether the problem is "too many poor people," "not enough jobs in the neighborhood for everyone who wants one," "too many government programs that destroy a will to work," or a host of other possibilities. All of these might be considered a "problem with poverty" in your community. Here is where it is useful to know the types of programs and activities that funders are interested in addressing now, so that you can respond with a plan to write a grant on some aspect for which resources are actually available.

Often, the literature can give you a precise definition of a problem. For example, many government programs are set up to assist people who are "poor." In this case, the term "poverty" has a precise meaning that you should know and use in your grantwriting.

Let's work through an example of how to understand the process of defining terms carefully within the context of grantwriting. To do so, we use funding opportunity announcement (FOA) HHS-2013-ACF-OCS-EE-0583, which was developed by the Department of Health and Human Services, Administration for Children and Families, Office of Community Services, Community Economic Development Projects (Department of Health and Human Services [DHHS], 2013). (You should download this document from http://www.acf.hhs.gov/grants/open/foa/files/HHS-2013-ACF-OCS-EE-0583_1.pdf. If you print it out you will be able to follow with the discussion in this chapter better.) While

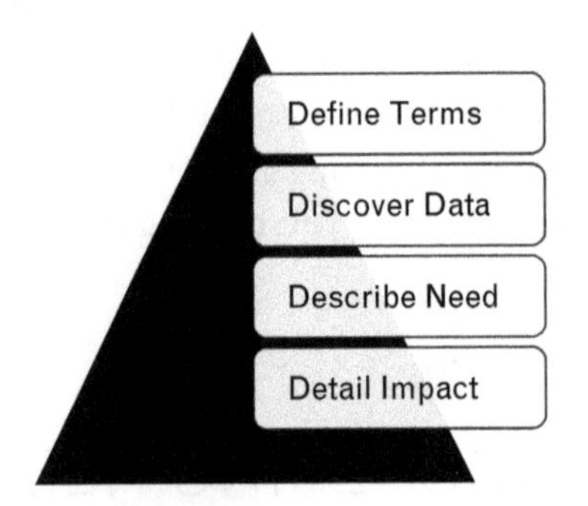

FIGURE 5.1: The 4 "Ds" of Uncovering Need

it is true that this example is from 2013, I am using it because the principles are timeless and I was an invited reviewer of the grant applications submitted under this FOA and thus judged how well people did with their applications. The FOA indicates that it "*seeks to fund projects that address the personal and community barriers that must be overcome to help low-income individuals become self-sufficient*" (p. 1, emphasis added).

This one sentence contains at least three terms that the grantwriter should be able to identify and define precisely. What do you think they are? Go ahead and write them down on a piece of paper. You can add the definitions next to them to begin to create your own grantwriting glossary.

The terms I see as being essential to understand are: (1) "personal and community barriers" (although this might be considered two different terms); (2) "low-income individuals"; and (3) "self-sufficient." What do you think these mean? Just for practice, write down what you believe is a reasonable definition of "personal and community barriers." Do the same with the terms "low-income individuals" and "self-sufficient." It is important that you actually write out definitions somewhere so that you can compare them to the ones given by the funder.

Because this is a well-established area, the grantwriter can look into the funding opportunity announcement (FOA) to find definitions of these terms. Let's examine "community barriers" first, as there is no separate definition of "personal barriers" under the "definition of terms" heading.

> **Community Barriers**: Conditions in a *community* that impede success in employment or *self-employment* of *low-income individuals*. Such conditions may include: lack of *employment education and training programs*, lack of public transportation, lack of markets, unavailability of financing, insurance or bonding; inadequate social services such as employment services, child care or job training; high incidence of crime; inadequate health care; or environmental hazards such as toxic dumpsites or leaking underground tanks. (DHHS, 2013, p. 4) [NOTE: italicized words or phrases are terms also defined in the FOA.]

As you wrote up your informal definition of this term, did you include all of these community barriers to self-sufficiency? As someone trained in social work, I included most of the social service-related items, but I missed the topics related to markets, financing, insurance, and bonding. How did you do?

In an interesting glitch, the authors of the FOA did not include a definition or examples of "personal barriers" with the definitions of all the other terms. The term *is* illustrated, however, in a different location: the section called "Application Review Criteria." Because the term is described under the criteria for reviewing applications, the term is vital to understand. This glitch, where an important element is hidden, emphasizes that grantwriters *must* read every word of the FOA before concluding that a particular bit of information is not included. In this case, the FOA states, "Examples of personal barriers include limited education, substance abuse, insufficient life skills, criminal history health problems or disability" (DHHS, 2013, p. 35).

Whenever you see examples such as these used to flesh out definitions of terms, be sure to highlight them. You will often want to begin to locate exact statistics for your community on these examples (this is explained more in the section "Step 2, Discover Data").

Let's do this again, with the term "low-income individual." What's your definition? Here's a clue—it is NOT the same as being poor.

Here's the official definition of "low-income individual": "An individual whose household income level does not exceed 125 percent of the official poverty guidelines as found in the most recent revision of the HHS Poverty Guidelines published by HHS" (DHHS, 2013, p. 6). Because these guidelines change every year, you need to keep looking up the exact dollar amount this refers to, but it is a clear number of dollars that makes a person low-income or not.

For the sake of completeness, let's conduct this definitional exercise once again. What does it mean to you to become "self-sufficient"? What change would a low-income person have to make in terms of income to become self-sufficient?

According to the Department of Health and Human Services, Office of Community Services, "self-sufficiency" means "a state of being or status of an individual or family where, by reason of employment, eligibility for public assistance is replaced by the financial capacity to meet all basic needs" (DHHS, 2013, p. 8).

These terms we have looked at lay the foundation for any grant application written in response to this announcement of a funding opportunity. The goal is laid out clearly in the context of these definitions. At its heart, you could paraphrase the purpose of this grant to "get people off of welfare by making it possible to get a job." There's a lot more nuance to the grant's purpose than that, but that is the essence. With these terms in mind, you can more easily make a decision at the start of the grantwriting process as to whether this is an opportunity too good to pass up, or outside the realm of what your organization does.

This definitional process is essential for analyzing every grant opportunity you find. If you don't have clear definitions for all essential terms, you won't know what you're talking about. It's like using a term from another language that someone tells you means something innocuous. It may actually be an insult, a curse word, or a vulgar term that undermines your credibility and standing instantly. Be sure to know what the words in a grant FOA or notification mean to the person reading your grant application. You can't just make up your own definitions and expect to be successful.

But what happens when the funder doesn't provide such clear definitions for the key terms? In this case, you still need to define your terms carefully and let the reader know where your definition comes from. Ideally, this will involve going to the research-based literature on the subject. For example, if you were addressing the issue of "depression" in middle-school-aged girls, you would use a definition from a medical or psychological assessment source, not just Webster's dictionary or some online, unattributed source.

A common language definition from the *A.D.A.M Medical Encyclopedi*a (2013) defines depression in this way: "Depression may be described as feeling sad, blue, unhappy, miserable, or down in the dumps. Most of us feel this way at one time or another for short periods." It goes on to indicate that clinical depression is an illness where these feelings last a long time and interfere with life activities. You would want to use the second discussion of depression as a clinically diagnosed illness rather than the first, more informal definition. By doing so, you appear more credible, scientifically oriented, and able to measure changes in the service recipients (a topic we return to when we discuss evaluation plans for your program).

STEP 2: DISCOVER DATA THAT SHOW THE EXTENT OF THE NEED IN YOUR COMMUNITY

When you use standardized definitions of terms you will find it much easier to find official statistics related to those terms. This is undoubtedly the best way to uncover need in your community to put into your grant applications. This section points you to some of the major government data sources.

The definitions of the HHS Office of Community Services are helpful now as well. In order to be funded from this FOA, applicants must provide "recent evidence (published within last 5 years) from U.S. Census updates and other statistics published by federal, state, county, city or other government bodies that both the unemployment rate and the poverty rate within the project's service area are equal to or greater than the state or national level" (DHHS, 2013, pp. 35–36).

If you were interested in finding this data for a grant application the first place to begin is with the United States Census Bureau (2017) (www.census.gov). Access is now available through a smartphone application that you can download too. The decennial census is taken every ten years as mandated in the Constitution and is the "gold standard" of data, despite some populations (such as the homeless) being undercounted. But the Census Bureau also operates between these big census dates, gathering information on a more frequent basis and making estimations based on actual data and projections. Every five years (those ending in 2 and 7), for example, an economic census is conducted that gathers information relating to business and the economy from the national to local levels. It is thus possible to find a great deal of information about social indicators of need that is relatively current. It is important to note that most data will be one to two years old, even though it is the most recent available.

One of the great benefits of using Census Bureau data is that it is frequently broken into smaller geographic areas. Good information is available at the county and often city level. Let's see how we could go from a national figure on poverty rate (as the FOA we are using for an example requires) to a smaller area. One quick way to do this for commonly used statistics is through the QuickFacts portal, which provides information "for all states and counties, and for cities and towns with more than 5,000 people" (see Figure 5.2.)

Looking at the Census Bureau's website (www.census.gov), you locate the QuickFacts portal in middle of the page (follow the arrow in Figure 5.2). Using the drop-down menu labeled "Select a state to begin," you can pick any state you like to begin your search for useful information. For this illustration, I have chosen Texas. When you choose a state from the drop-down menu, QuickFacts will automatically load the page for that state.

The website shows what you can see in Figure 5.3. This provides an overview of many facts about Texas. But we wish to find comparative data, so we look at the box that Arrow A points to that says "enter state, county, city, town or zip code." We enter "United States" first and hit return to add national data to the display. Then we enter Dallas County, Texas" to show data from that county in addition to US and Texas information. Finally, we go the "All Topics" box (arrow B) and search for the label "income and poverty." When we've done that small level of manipulation of the site, we have what is shown in Figure 5.4, a list of data from the four levels of jurisdiction: national, state, county, and city.

Here you can see information about the percent of persons in poverty (which is the information requested in the grant application). While the Census Bureau cautions about

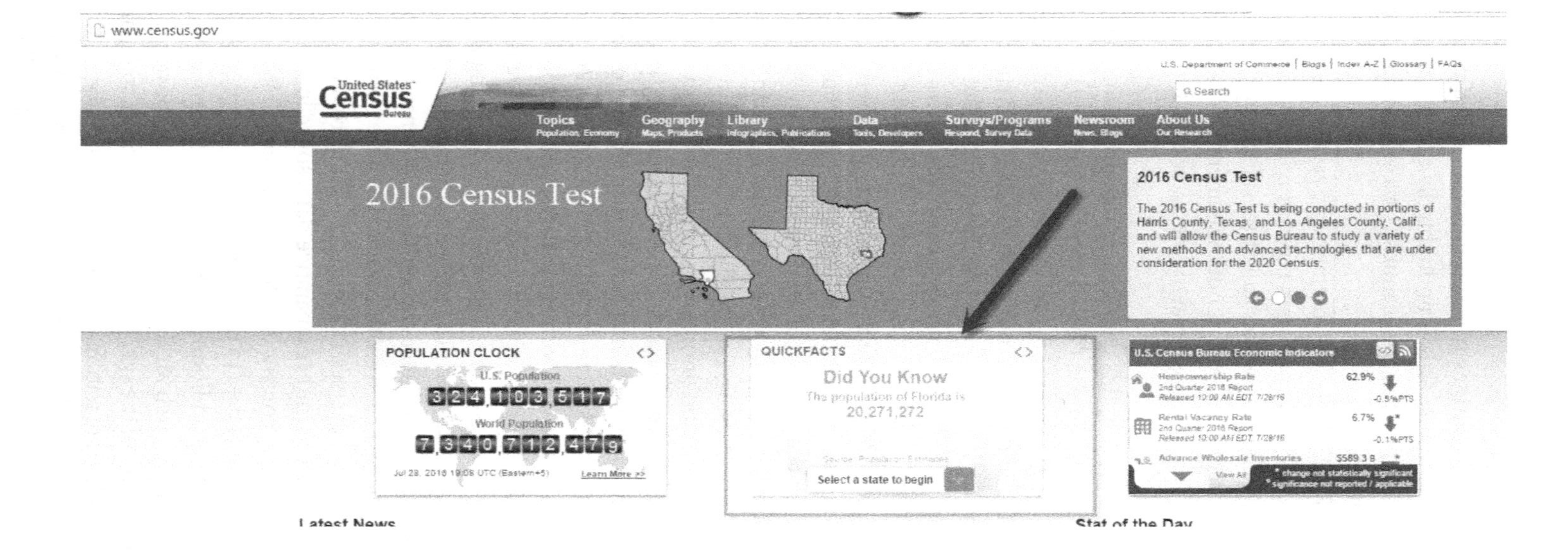

FIGURE 5.2: Finding the Census Bureau's QuickFacts Portal

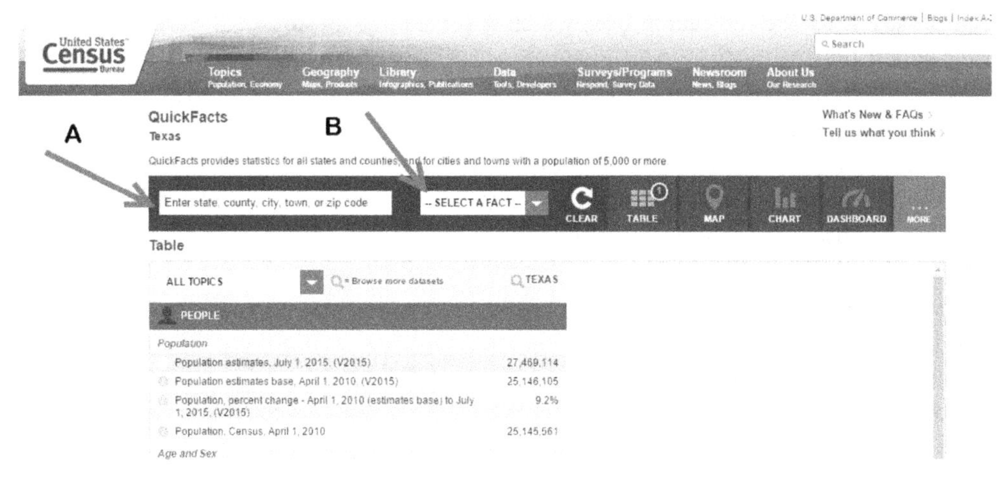

FIGURE 5.3: Texas's QuickFacts Page

comparing poverty data from smaller areas to other areas, it is still possible to see a pattern in the information that highlights the facts that these jurisdictions are worse off than the national average. The average median household income in Texas ($52,576) is lower than in the United States as a whole ($53,482). Similarly, median household income in Dallas County ($49,925) is lower than in Texas. The City of Dallas is lower than the county ($43,359). Poverty in Texas, at 17.2 percent, is higher than at the national level (which is 14.8%). Dallas County's poverty rate is higher than the state of Texas, at 19.3 percent. The City of Dallas has a poverty rate of 24.1 percent, the highest of all jurisdictions. Using this string of figures, we can begin to show that there is greater need in Dallas than other parts of the United States and even in other areas of Texas. This information thus begins to paint a picture of need for an antipoverty effort in Dallas. (You can drill down in the information to the city level, as long as cities or towns have a population larger than 5,000 people.) Because of the ease of use of the QuickFacts site, discovering this data is very fast, once you know how to do it.

Not every bit of information you will want is available in the QuickFacts portal. You will also need to become familiar with other pages on the Census Bureau website. One way to gather information quickly when you don't know where to locate it in the Census

Enter state, county, city, town, or zip code | -- SELECT A FACT -- | CLEAR | TABLE | MAP | CHART | DASHBOARD | MORE

Table

INCOME AND POVERTY	DALLAS CITY, TEXAS	DALLAS COUNTY, TEXAS	UNITED STATES	TEXAS
PEOPLE				
Income and Poverty				
Median household income (in 2014 dollars), 2010-2014	$43,359	$49,925	$53,482	$52,576
Per capita income in past 12 months (in 2014 dollars), 2010-2014	$27,917	$27,195	$28,555	$26,513
Persons in poverty, percent	24.1%	19.3%	14.8%	17.2%

FIGURE 5.4: Census QuickFacts Page Showing U.S., Texas, and Dallas County Data

Bureau site is to use the search box, which is located at the top of the website on the right side.

Grantwriters can use this tool to discover the unemployment rate for the United States, Texas, Dallas County, and City of Dallas. Using information from the American Community Survey for 2012, we see that the United States' unemployment rate was 9.4 percent, Texas' rate was 7.7 percent, Dallas County's rate was 8.7 percent, and City of Dallas' rate was 9.0 percent. Discovering this information took very little time, and it is exactly what is being asked for by the FOA. This information does not actually create a picture of strong need for the program goals set forth by the Office of Economic Services, since the unemployment rate in the City of Dallas is a bit better than in the country as a whole, but the city of Dallas's numbers are worse than in the county and the state. Relatively speaking, there is need on this one indicator as defined in the FOA and the requirements set forth for unemployment are satisfied.

Of course, not all information is going to be available from any one source. As a grantwriter needing specific indicators of need, you may have to develop good relationships with researchers in various government offices at the state, county, city, school district, and other levels in order to get that one elusive bit of information that will make your case as strong as possible.

Information that is useful to uncover need can also be found in other venues. Legislative hearings, community needs assessments conducted by a nonprofit organization such as a local United Way, newspaper or magazine articles, conversations in hallways at meetings of nonprofit leaders—all these can help to open your eyes to needs that might be addressed. It is up to you to be ready to dig deeper for reliable information to support a diagnosis of a true need in your community. With knowledge of the current grant opportunities available, you already know which needs you can access resources for, and which you will find difficult or impossible to acquire funding for.

Once the information has been discovered, it is time to put it together into a coherent whole. That is the process completed in Step 3.

STEP 3: DESCRIBE THE NEED IN YOUR COMMUNITY, POSSIBLY COMPARED WITH OTHER COMMUNITIES

There is an important difference between Step 2, discovering data about a need, and Step 3, describing the need. It is the difference between a "mere collection of facts" and a "compelling story" about your community. Think about this difference carefully. Facts are often derided as "factoids" because they are small in scope, isolated, and unconvincing on their own. The key to turning a set of individual facts into a story is adding context.

At the risk of oversimplifying this idea, let's consider the difference between a boring person at a party and someone whom other guests crowd around. A boring person generally talks about himself and what's important to him. His conversation is sprinkled with facts that don't connect to each other or to listeners. The stories a bore tells frequently are not arranged in chronological or other order—no thought is given to the context, and the bore will stop to tell you something that was important but not mentioned, then try to start up the narrative again. Wording is haphazard, with little effort made to involve the senses or imagination of the audience. Finally, when the story is completed, listeners might be left wondering what the point of it was. Sound familiar?

Contrast this to the "life of the party"—the person who has an audience consistently through the evening. What does she do? One reason people want to listen is because the stories told relate to them, not just the teller. Even if the story is about an extraordinary journey that no one else in the room has taken, somehow she is able to bring us along so that it seems possible that we, too, were there. This storyteller provides us the information we need to stay in the narrative, at just the time we need it. The information we receive fits together, each bit connecting with what has come before and what will be presented later. If a sudden change in direction is needed in the story, this person signals the shift, not leaving the audience to wonder why or how it is important. Most importantly, at the end, the listeners feel that they have been given a complete story, that they have learned something valuable, and that their time was well spent.

Your job, in uncovering needs, is to tell a great story about why your community, organization, or group has reason to seek assistance to solve a clear problem. Your job, at this stage, is to get the reader to look up from the text and say, "This situation is truly intolerable."

The way to do this is to master the difference between reciting facts (such as unemployment and poverty rates) and truly describing the needs of your community in the context of the goals of the funder. It is one thing to write that your community has a high poverty rate, which is a problem. It is another thing, and more effective in the case example we are using in this chapter, to describe the high poverty rate as a result of a lack of educational and employment opportunities. Connecting the basic fact of having many community members being poor to the facts of high numbers of school dropouts, or a dearth of jobs in the community within easy reach of residents, provides the reader with an idea why a problem exists. This context allows readers to understand the "why" as well as the "what."

A very useful tool in understanding connections and context is to create a concept map. Originally developed to help understand children's knowledge of scientific ideas (Novak & Canus, 2006), a concept map is a graphical representation of connections and links between ideas. Words describing specific concepts are usually placed in circles or boxes, and lines are drawn between the concepts that are related. It can be helpful to write one or a few words next to the connecting line to describe the nature of the relationship between the two concepts. Because of the contextual nature of how concepts are connected, it is helpful to have a focusing question. Figure 5.5 shows a concept map that relates to the FOA example we have been using in this chapter.

This is not a complicated example, but it clearly shows the funder's goal of economic development as a catalyst for individual self-sufficiency. Also shown are both community barriers and personal barriers that block the attainment of individual self-sufficiency. The four community barriers were identified from among those listed by the funder as being applicable to the community, while the three personal barriers were also identified as being important as blocks toward self-sufficiency by the Office of Community Services.

As laid out here, one can imagine finding data to support all or almost all of these barriers being present in a community. Information about (a lack of) employment opportunities could be found in the Census Bureau's surveys of economic activity. A community survey done by United Way, or even an examination of a phone book, might disclose the number of

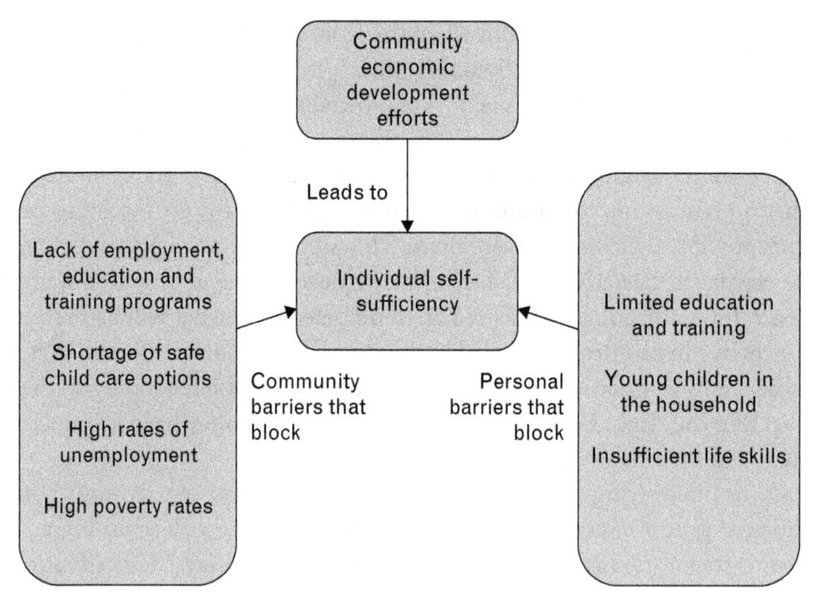

FIGURE 5.5: Self-Sufficiency FOA Concept Map

training programs and licensed child care centers in the community. Also from the Census Bureau, statistics on unemployment and poverty can be found quickly.

Information on personal barriers might not be as easy to find, but it is probably available. The Census Bureau has information about educational levels, as well as the number of children below the age of 5 years at the community level. Making a case that residents in a community have insufficient life skills may prove a challenge unless there has been a special survey taken on the topic in your community. Still, working with local schools, churches, or other institutions may turn up evidence along this line.

The concept map has been introduced as a way to systematize the variables being examined and their relationships to each other. The concepts in the example could probably also be further linked to each other, such as the concepts of lack of education and training programs, unemployment, and poverty. Even a basic concept map such as this one is able to move the grantwriting process along quickly. A concept map also serves as a graphic reminder of the main ideas being worked on by the funder and the applicant. The data that need to be collected and connected are also quickly referred to.

The facts of high poverty and unemployment rates mean something to the Office of Economic Services (the funder) only in terms of its goals with this FOA to address personal and community barriers that prevent self-sufficiency. The funder wants to target communities where the problems of poverty and unemployment are higher than average in order to help those who are needier, but funds exist to assist only a very few projects. It is thus vital to link the story to the goals of the audience, that is, the funder.

Because of the desire to make a mark on the problem by targeting the areas of greatest need, compelling descriptions of need often compare two locales with one another, if only to show the need is relatively higher in your community than elsewhere. This has already been demonstrated by the requirement in the FOA to ensure that unemployment and poverty rates are at least as high in your locale as in your county, city, or state.

An excellent statement of need also does something else that marks the difference between a decent write-up and an excellent one. It looks forward to other parts of the grant application. This is done in two ways. First, an excellent statement connects the problem being discussed in this application to the types of problems that the grant-seeking organization typically addresses. This is important because the funder wants to give resources to organizations with some expertise and experience in effectively addressing the problem at hand.

The trick at this stage is to give the reader knowledge about the problem that will later be connected to the organization's past and current capacity. While federal funders do not always explicitly require proof of success with past efforts, such history strongly suggests that the agency has the capacity to be successful in the future. This particular FOA, for example, states that applicants and/or their partners must provide evidence of successful completion of at least two projects similar to the one proposed.

The other attribute of an excellent need statement that sets it apart from others is that it provides foreshadowing of the solution that will be proposed to address the problem (we discuss the links between how a problem is described and the solutions that are set forth in the next chapter). The way you describe the need and its causes implicitly leads to an understanding of how to solve the issues you present.

For example, if high poverty and unemployment are described as being linked in your community to having few opportunities for entrepreneurial activities, the solution proposed later should have a component to increase opportunities for entrepreneurial action. If the reader can't connect the need with the proposed solution, funding is unlikely. Making this connection is made easier if you use a logic model, a graphic that links the problem to resources used, outputs achieved, and outcomes accomplished. We'll cover how to create a logic model later.

STEP 4: DETAIL THE IMPACT OF THE NEED OR PROBLEM ON PEOPLE IN YOUR COMMUNITY

The final element of uncovering need is to go beyond the numbers to detail the impact of the need on people in your community. This is not the time to go into a comprehensive case study of a random set of individuals. It is, however, important to put a human face to the problem so that readers connect on an emotional level as well as a rational level.

Knowing human physiology is helpful here. We have in our brains a set of cells known as "mirror neurons." Some researchers believe that mirror neurons in humans are associated with empathy and the ability to understand other people's motives. Painting a verbal picture of others' actions can, they assert, trigger a response of empathy on the part of the reader (Winerman, 2005). If this is true, evoking such a reaction through detailing the impact of the problem on people in your community may trigger a higher emotional connection with the grant proposal, leading to the possibility of receiving a higher score when it is assessed. While this is a controversial aspect of both brain chemistry and grantwriting, there is no doubt that clear writing filled with accurate, interesting, context-laden prose keeps the reader's attention, creating a better chance for thoughtful critique and judgment on the application's merits.

Let's try our hand at this. For this grant FOA, examples of "community barriers" have been named, such as inadequate child care or job training. Using fictitious numbers for this example, we might establish that these barriers exist in your community by citing official statistics showing that:

> There are only five licensed day care facilities in our community of 14,765 people. Altogether, they have a total capacity of about 200 preschool aged children. Yet, there are over two thousand children (2,085, or 14% of the community) under the age of 5. This amounts to only one licensed day care facility per every 417 children and leaves most young children (1,885) with their mothers (predominantly) or with unlicensed day care providers. People with young children thus face a significant barrier to becoming self-sufficient.

This is not a bad statement of the problem, and it sounds good at first. But it doesn't actually link to the funder's goal of creating self-sufficiency or involve the reader emotionally. Here is an effort to improve it.

> Parents in our community face a frustrating barrier in their quest to become self-sufficient while providing safe care for their young children. Only five licensed day care providers exist for the over 2,000 children in the community under the age of 5, leaving the other 1,800 children to be cared for by their parents or unlicensed care providers. This means that a large number of parents are not able to work due to a lack of child care, blocking their path to moving off of public assistance.

This revised version is almost exactly the same number of words long but brings more emotion to the writing by focusing on the frustration of parents who want to become self-sufficient but have nowhere to leave their children. A few words have been selected specifically for their emotional impact—"frustrating barrier," "providing safe care for their children," and "blocking their path." The facts provided are generally the same but the emotional pull is greater in the second effort.

In both versions, you can easily imagine that the authors of this (fictitious) needs statement may be laying the groundwork for their solution, starting additional licensed day care centers. Such a plan would both eliminate a barrier to self-sufficiency and create new jobs as child care workers.

The key to detailing the impact of the problem is to show how it affects people and to link it very explicitly with the funder's intentions and goals. Connecting with the funder's stated desires through the facts and emotions you evoke are vital characteristics of successful grantwriting in the Age of Scarcity.

Grantwriters are almost always up against tight deadlines and space limitations. It is thus important to write with the minimum number of words possible. At the same time, you must connect with readers on a human level so that reviewers can imagine how the funds will address an important need. The funding to be asked for will improve the lives of real people, not impersonal abstractions. If you don't provide the reader with enough details to share an emotional link with people in your community, the chance of being funded is decreased. This may seem an impossible task, to be numbers-based AND emotionally connected, but with practice and focus you can do it.

SUMMARY

This chapter lays out a framework for writing about needs in your community. The core message to recall is the "4 Ds of Uncovering Need." These are the four steps that will take you far in bringing to light why funding is needed. In short, the steps are:

1. Define the terms;
2. Discover the data;
3. Describe the need; and
4. Detail the impact.

The chapter provides additional information for each step. One of the most important elements of this approach is to connect the desires of the funder with the needs of your community. Once you define the terms that are important in the FOA or request for proposals (RFP) in the way the funder does or in a way that is consistent with the academic literature, you will be able to discover data to describe the need in your community. The final step is to detail the impact of the need or problem on the people in your community. This places you in a very good position as you move into the next part of the grant proposal, providing a solution.

PRACTICE WHAT YOU'VE LEARNED

Use an FOA or RFP or funding opportunity from a foundation website to complete the following steps.

STEP 1: DEFINE THE TERMS

1. Read through the document carefully, from start to end. Highlight key terms that may be defined for you or not, but that seem key to understand. It is helpful to start a list of key terms in a separate document.
2. Determine whether the FOA document has a definition for or examples of each of these key terms. Sometimes, as in the example used in this chapter, the definitions or examples of what terms mean are not where you expect them to be. Transfer the definitions or examples to your list of key terms.
3. For all remaining key terms (if any), search the literature for official government definitions or at least what the research community would consider a valid definition or examples of the term. Write these on your list of key terms. You have now completed Step 1 of the 4 Ds of Uncovering Need, "Define the Terms."

STEP 2: DISCOVER THE DATA

1. For each of the terms you defined in Step 1, search for a data source to determine how prevalent it is in your community. For example, if one of the key terms in the FOA you are using is "poverty," find out from official sources the extent to which poverty impacts your community. Some terms and ideas may be more difficult to locate. Keep track of which terms you have information on and which you do not.
2. Looking now at the terms for which you cannot find data (if any), develop a strategy to be able to talk about these concepts in the context of your community. Look outside the usual sources for information—this is your chance to be an information detective. You flagged these terms as being important when you read through the FOA, so you'll want to determine a way to discuss them in your grant application. Now that you have discovered the data, it is time for Step 3.

STEP 3: DESCRIBE THE NEED

1. Read through the main findings you have gathered so far. Connect them in your mind, using a technique such as concept mapping, to show linkages between them. Which data measure concepts that cause, are related to, or flow from other concepts? Create a diagram to show these relationships.

2. You've now developed some context for the bare-bones data. This will help you write up the need in your community as a result of other factors (perhaps ones that your proposal will address) and also lead into Step 4, where you detail impacts on clients.
3. As you work on your need description, remember to connect your write-up to the desires of the funder. If you want the funder to provide you with resources, you will be required to show how addressing your community's need is related to achieving the funder's goals.

STEP 4: DETAIL THE IMPACT ON PEOPLE

1. You've done a lot of work so far on uncovering need. This last step is to remind you to link the funder's goals, your community's context, and the relevant data you've gathered back to the lives of the people in your community. Think about how the need you've uncovered causes problems for actual individuals. Are they unable to obtain employment? Do they suffer from poorer health than they should have to? Are children or the elderly put into danger? If you're not sure what to write here, get out of your office and talk with front-line workers in your organization, service recipients, or members of your community who may have more direct knowledge of the problem's impact than you have.
2. After you've determined what the impacts are, revise your written statement of need to incorporate your new insights. Show it to others who have not been involved in writing it. Note their reactions. You want the reader both to have a strong "feeling" about the impact of this problem on your community *and* to be convinced that your facts support the importance of the issue.

REFERENCES

A.D.A.M Medical Encyclopedia (2013). Major Depression. Retrieved from http://www.ncbi.nlm.nih.gov/pubmedhealth/PMH0001941/#adam_000945.disease.causes

Department of Health and Human Services (2013). HHS-2013-ACF-OCS-EE-0583. Retrieved from http://www.acf.hhs.gov/grants/open/foa/files/HHS-2013-ACF-OCS-EE-0583_1.pdf

Novak J. & Canus, A. (2006). Theory Underlying Concept Maps. Retrieved from http://cmap.ihmc.us/Publications/ResearchPapers/TheoryCmaps/TheoryUnderlyingConceptMaps.htm

United States Census Bureau (2016). Home. Retrieved from www.census.gov

Winerman, L. (2005). The Mind's Mirror. American Psychological Association. Retrieved from http://www.apa.org/monitor/oct05/mirror.aspx

CHAPTER 6

FINDING AND CREATING EVIDENCE-BASED PROGRAMS

The implicit promise that you make as you write about the problems in your community is that you will do your best to remedy them. This is the contract that you, as the funding seeker, make with the organization that will provide you with resources. Just as you have done a good deal of research to understand and be able to explain the problems facing your community, you must also understand in some detail the solutions that you present to the funder.

There is no doubt that you want to make the situation you find in your community better. And there is no doubt that the funder wants the same thing. It doesn't matter if you have a private foundation or a government agency as your backer—you and they want to solve problems. The key question for both you and the funder is how to impact the problem in an effective way. It is your job in the grant proposal to show that the way you want to address the problems mentioned will work.

Given that you probably haven't actually implemented the project yet, this seems like a tall order. It can be, but there is one convincing way to make a case that your approach is more likely to succeed than not: Use program ideas that have already been determined to be "evidence-based" successes.

WHAT DOES "EVIDENCE-BASED PROGRAM" MEAN?

The term "evidence-based program" (EBP) refers to a program that has been designed to solve a particular problem and that has been proven through research to reduce the problem in the desired way. As one arm of the US federal government has stated, "An evidence-based program (EBP) is a program proven through rigorous evaluation to be effective" (Family and Youth Services Bureau, n.d., p. 1).

This concept emerges from evidence-based practice, and derives from the term used in the field of medicine to denote practices that are effective in decreasing particular medical issues. In medicine, of course, double-blind tests of new drugs and procedures are standard practice and the efficacy and safety of anything new needs to be proven through clinical trials before being released to the public. While the same level of proof is not usually available for social program interventions, this is the model on which evidence-based practice in social work and human services in general is based.

New research is constantly being released through conferences and academic journals. It takes a lot of effort to stay current with what the latest evidence-based practices are. So many studies are completed that it is difficult to be sure that a program backed by solid evidence a few years ago is still seen as more effective than more recently developed programs and practices. Thus, being an evidence-based practitioner in a program based on the best evidence available is an ongoing process (see Gambrill, 1990, 1999; Gibbs & Gambrill, 2002). One of the earliest writings on EBPs (Sackett, Richardson, Rosenberg, and Haynes, 1997, p. 3) also argues that evidence-based practice is a problem-solving process consisting of five steps:

1. Convert information needs into answerable questions.
2. Track down, with maximum efficiency, the best evidence with which to answer these questions.
3. Critically appraise the evidence for its validity and usefulness.
4. Apply the results of this appraisal to policy/practice decisions.
5. Evaluate the outcome.

While these steps make a lot of sense when dealing with medical or direct clinical practice, Hoefer and Jordan (2008) argue that some aspects are missing in Sackett et al.'s process if you are a person designing a program to address social problems at a macro level. The first is that identified needs must be at a community or societal level. For example, a good clinical-level question might be "What is the most effective way to treat depression among African American youth?" For a grant proposal operating at a community level, however, a more relevant question might be "What is the best way to reduce levels of depression among African American youth in our community?" It may be that training additional psychotherapists or social workers in effective treatment modalities is the best way. Or it may be that changing community-level conditions such as high unemployment and low educational levels is more effective.

Steps 2 and 3 are similar in program development, but Hoefer and Jordan (2008, p. 551) suggest a new step 4: "Provide clients with appropriate information about the efficacy of different interventions and collaborate with them in making the final decision in selecting the best practice." The client, in this case, may be representatives of the target population, an advisory group working with your agency on a continual or ad hoc basis, or a coalition of interested agencies willing to collaborate to address the problem. The point of this newly inserted step is to receive input from the affected population and to ensure that proposed solutions are culturally relevant and acceptable.

Additional steps from Hoefer and Jordan (2008) include:

5. Apply the results of this appraisal in making policy/practice decisions that affect organizational and/or community change.

6. Assess the fidelity of the implementation of the macro practice intervention.
7. Evaluate service outcomes from implementing the best practice.

These later steps are covered in future sections. For now, however, the key idea is that finding the "best" program to solve the identified problems may require considerable research and continued commitment to updating your knowledge about what works best to address the problems that your organization wants to tackle.

WHAT MAKES UP AN EVIDENCE-BASED PROGRAM?

All programs are made up of various components that come together to try to improve lives. In programs that are *not* evidence-based, aspects or parts of the program are included because the designer "feels" or "believes" that they are important. Evidence-based programs must meet a higher standard with all of the parts of the program being tested to determine whether they help achieve program outcomes. Evidence-based programs usually have two types of components: core and "adaptable." Core components are based on theory and need to be implemented according to the program developer's guidelines if you wish to achieve the results that have been shown to result from the program. According to the Substance Abuse and Mental Health Services Administration (SAMHSA, n.d.), core components are made up of the following:

- Staff selection;
- Pre-service and in-service training;
- Ongoing consultation and coaching;
- Staff and program evaluation;
- Facilitative administrative support; and
- Systems interventions.

The Family and Youth Services Bureau of the US Department of Human Services indicates that core components (for educational programs) fit into three categories: content, pedagogy, and implementation. Thus important core components always include *what* is taught, *how* it is taught, and the *logistics* that go into creating an environment conducive to learning (Family and Youth Services Bureau, n.d.).

While "core" components are those that may not be changed, lest the efficacy of the intervention be lost, other aspects of the planned program may be considered "adaptable" because they can be changed to meet the needs of another locale, population, or culture. Adaptable components are ones that have been tested but are not necessarily vital to achieving positive results. Choosing a program based on the evidence that it is effective but then changing the core components of the program is quite unsound and can result in a waste of considerable time, energy, and other resources. Altering "non-core" or adaptable components is a reasonable effort to increase the "fit" of the evidence-based program to a new environment and situation.

An example of the difference may be helpful. A core component of the CAST (Coping and Support Training) program (which is an evidence-based program that we'll encounter again later in this chapter) is that twelve 55-minute sessions are held over a 6-week period, covering specific topics in a specific way as shown in in the course curriculum notebooks (National Registry of Evidence-Based Programs and Practices [NREPP], 2014). If an organization changes the number, length, timing, or content of the sessions, it will be altering at least one core component. On the other hand, the age of the students may be adaptable. Currently, the information about the program indicates that positive effects are found when using the curriculum and program with adolescents and young adults, from the age of 13 to 25. Would this approach be as effective with "emerging adults" up to the age of 30? No information is currently available, but a case might be made that this is a reasonable extension of the initial research and so the age of the student could be seen as an adaptable component.

The Office of Justice Programs also has information about evidence-based programs:

> OJP considers programs and practices to be evidence-based when their effectiveness has been demonstrated by causal evidence, generally obtained through one or more outcome evaluations. Causal evidence documents a relationship between an activity or intervention (including technology) and its intended outcome, including measuring the direction and size of a change, and the extent to which a change may be attributed to the activity or intervention.
>
> Causal evidence depends on the use of scientific methods to rule out, to the extent possible, alternative explanations for the documented change. The strength of causal evidence, based on the factors described above, will influence the degree to which OJP considers a program or practice to be evidence-based. (Office of Justice Programs, 2014, p. 9)

FINDING EVIDENCE-BASED PROGRAMS

Certain government agencies have lists of programs that they believe have evidence to support their efficacy when implemented according to guidelines. These lists do change, as new programs are added that have gained research support and old, "promising" programs are dropped as new evidence determines the programs are not effective, after all. The rest of this chapter provides resources to find programs that have evidence to support their effectiveness.

THE NATIONAL REGISTRY OF EVIDENCE-BASED PROGRAMS AND PRACTICES

The NREPP is described on its website's home page (www.nrepp.samhsa.gov) as "a searchable online registry of more than 350 substance abuse and mental health interventions. NREPP was developed to help the public learn more about evidence-based interventions that are available for implementation. NREPP does not endorse or approve interventions" (see arrow A in Figure 6.1). The latest redesign of the NREPP website at the time of this writing was in September 2015. It may be revised by the time you read this chapter, but the basic functions and purposes will be the same.

The NREPP is not the end-all list of all programs that you may wish to be familiar with as you write grants in these areas. Still, it is a good place to begin a search for a program that your target funder will provide resources for because there is evidence that it works. The NREPP website is also a source for considerable detail to put in your proposal if you choose one of the programs listed.

Two other areas shown in Figure 6.1 are notable. The first, pointed at by arrow B, is an opt-in box where you can enter your e-mail address. When you do so, you'll be sent an e-mail to confirm that you wish to be on the NREPP mailing list to receive a monthly notification of NREPP news. This is highly recommended so that you, as a grantwriter, will receive information automatically on new programs that have been added to the list and old programs that may be taken off. You can place these e-mails into a "reminder" file on your computer or just use them as a way to recall that this resource exists. Grantwriters need this sort of idea generator at times.

The last area of note in Figure 6.1 is pointed to by arrow C. It is labeled "Find an Intervention" and is the heart of NREPP for grantwriters. This is the entry point into the database of interventions. When you click on this area, you are taken to another web page as shown in Figure 6.2.

FIGURE 6.1: NREPP's Homepage

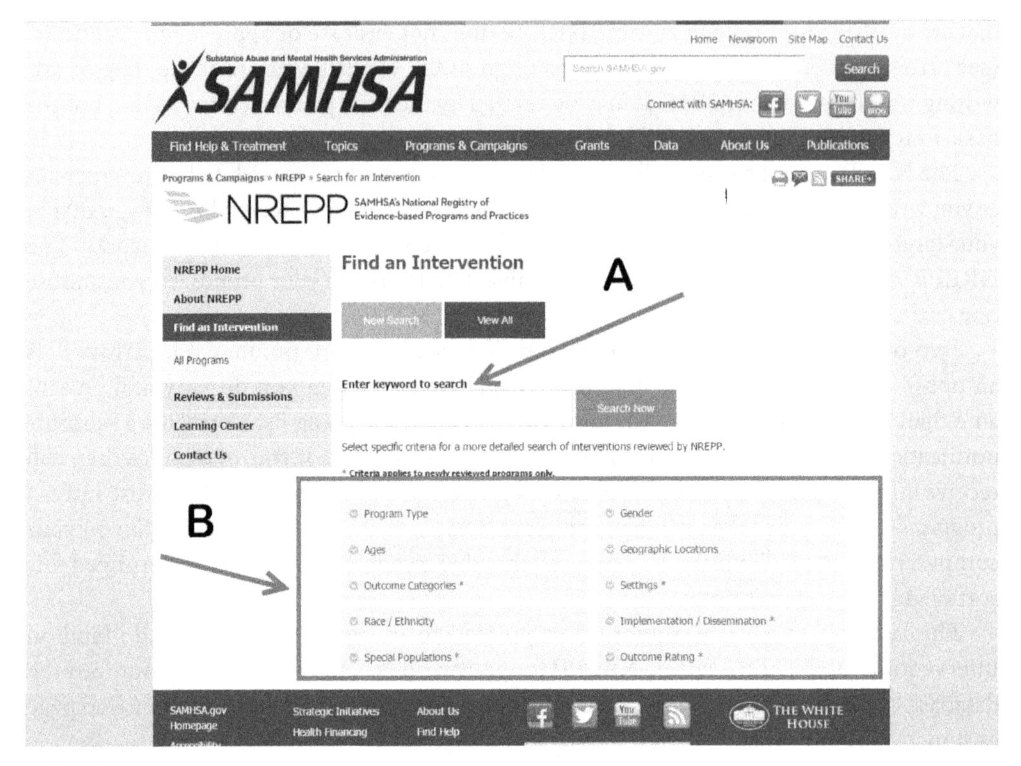

FIGURE 6.2: NREPP "Find an Intervention" Search Page

Arrow A in Figure 6.2 points to the main search box, where you can enter any substance abuse or any mental health treatment or prevention term. You can enter terms such as "opioids" or "heroin" relating to substance abuse treatment or prevention; "depression" and "PTSD" are examples of search terms related to mental health treatment or prevention. You may put in just the basic search term to see what comes up.

The NREPP site underwent a significant updating in September 2015. All the programs that were in the dataset at that time are now called "legacy" programs. The definitions of effective, promising, and ineffective programs changed as well. As you use NREPP, be sure to note that there are two datasets: "legacy" and "newly reviewed" programs. As you search, you will need to look first at one, and then the other.

The boxed area pointed to by Arrow B allows you to narrow the scope of your search immensely. Ten ways are presented to decrease your search's breadth—program type, ages, outcome categories, race/ethnicity, special population, gender, geographic location, settings, implementation/dissemination, and outcome rating. Table 6.1 shows these categories and the options you may select within each category. As you can see, you could search for an effective outpatient substance abuse treatment intervention that is geared toward Black, homeless, transgender adolescents living in urban areas, and that has implementation materials available. THAT would be very specific! We'll work through an example using NREPP, using the topic introduced earlier, depression among African American youth.

TABLE 6.1: Ten Categories of Search Terms and Their Options in NREPP

Major Category	Options Within Category
Program Type	• Mental health promotion • Mental health treatment • Substance abuse prevention • Substance abuse treatment • Co-occurring disorders
Ages	• 0–5 (Early childhood) • 6–12 (Childhood) • 13–17 (Adolescent) • 18–25 (Young adult) • 26–55 (Adult) • 55 + (Older adult) • Information not provided
Outcome Categories	• Mental Health • Substance Use • Wellness
Race Ethnicity	• American Indian or Alaska Native • Asian or Pacific Islander • Black or African American • Hispanic or Latino • White • Other • Information not provided
Special Populations	• Co-occurring Disorders • Couples • Families • Homeless or Runaway • Immigrant/refugee • Justice-Involved Adults • Justice-Involved Youth • Lesbian, Gay, Bisexual, Transgender, Questioning (LGBTQ)/ITS (Intersexual) • Low-Income • Military or Veteran • Non-English Speaking • Populations Affected by Emotional Disturbance (ED) • Suicidal • Transition-Aged Youth • Tribal or American Indian or Alaska Native • Victims of Trauma or Violence • Youth in or Transitioning out of Foster Care • In-Home Language Use (other than English) • Populations affected by Serious Mental Illness (SMI) • Older Adults
Gender	• Male • Female • Transgender • Information not provided
Geographic Locations	• Urban • Suburban • Rural and/or frontier • Tribal • Non-U.S. • Information not provided.

(continued)

TABLE 6.1: Continued

Major Category	Options Within Category
Settings	• Hospital/Medical Center • Residential Facility • Outpatient Facility • Correctional Facility • Court • Home • School/ Classroom • University • Workplace • Mental Health Treatment Center • Substance Abuse Treatment Center • Other • Information not provided
Implementation/Dissemination	• Implementation materials available • Dissemination materials available
Outcome Rating	• Effective • Promising • Ineffective

One of the tricks of any search system is to balance specificity with generality (think of this as the "Goldilocks Rule," where things need to be neither "too much" nor "too little" but "just right!"). Being either too specific or too general an approach can cause problems. Still, the NREPP database can accommodate a number of parameters and find results. To show what can occur, "depression" was entered in the NREPP search box with additional parameter choices of "adolescent," "Black," "male," "urban," and "school/classroom" for setting. Despite having five parameters limiting interventions regarding "depression," NREPP found two newly reviewed programs (but no legacy programs). (See Figure 6.3, arrow A).

Arrow B of Figure 6.3 points to a box showing that two programs were found in the search results. One program showed at least one "effective" outcome, both programs had at least one "promising" outcome, and both programs also showed at least one "ineffective" outcome. Arrows C and D lead to the title of the two programs, "Problem Solving Therapy," and "SOS Signs of Suicide: Middle School and High School Prevention Program," respectively. The next column to the right ("evidence rating by outcome") shows a number of possible outcomes that the programs address. Each one has a rating to show whether it has been shown to be an effective program to produce that particular outcome. In this case, the Problem-Solving Therapy intervention has evidence to support it being effective against "depression and depressive symptoms" and "suicidal thoughts and behaviors," as well as helping to achieve "personal resilience and self-concept." The Problem Solving Therapy program has some outcome results that are seen as "promising" (i.e., improving "social functioning/competence," increasing "self-regulation," and reducing "non-specific mental health disorders and symptoms"). The program is rated as being "ineffective" in achieving three potential outcomes: general functioning and well-being; phobia, panic, and generalized anxiety disorders; and physical health conditions and symptoms.

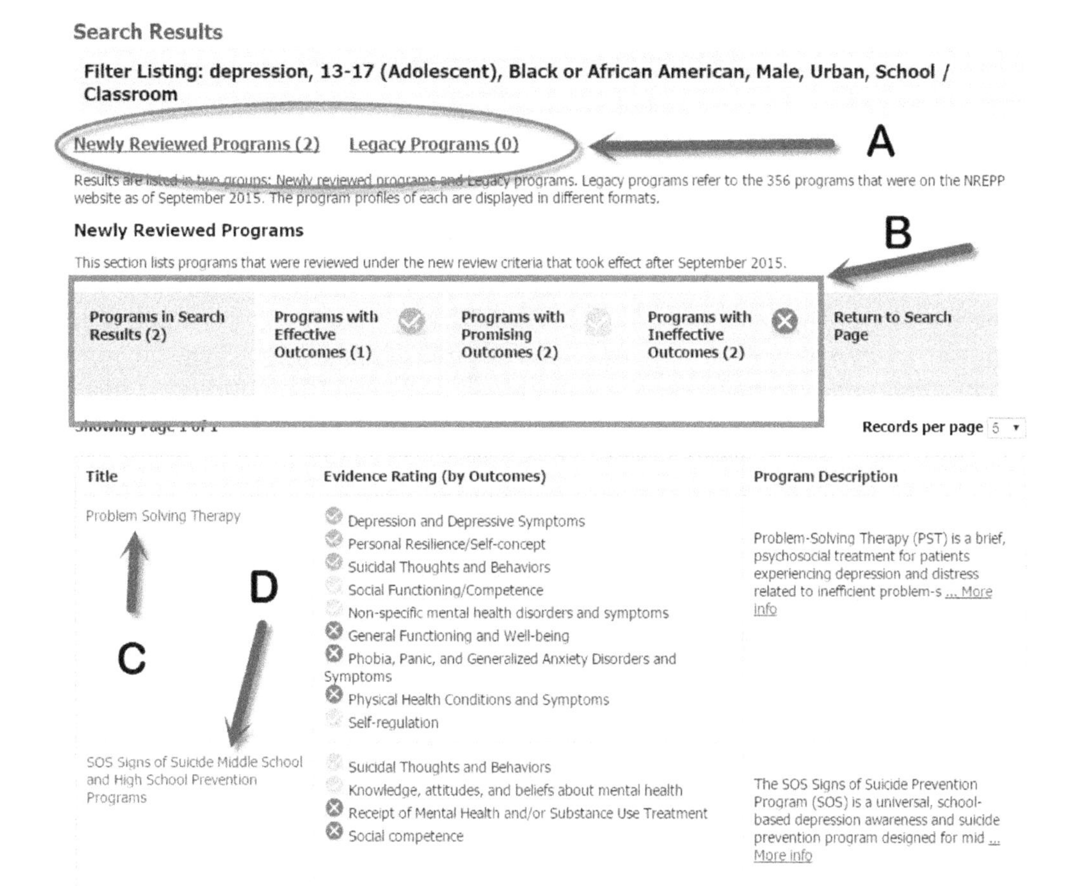

FIGURE 6.3: Search Results Using NREPP

The SOS program (arrow D) has no outcomes that are "effective." The program does have evidence to support that it may counter "suicidal thoughts and behaviors," and may positively alter "knowledge, attitudes, and beliefs about mental health." The SOS program is considered "ineffective" to reduce the "receipt of mental health and/or substance use treatment." It also does not impact "social competence."

Before moving forward with choosing either program, or searching for additional options, we can see the usefulness of this information. Depending on the desired outcomes for the intervention (which is related to how the community program has been defined), a grantwriter can quickly narrow down programs to those addressing successfully the needs earlier identified in the grant proposal. If the need is related to high levels of depression or depressive symptoms, the first program has been shown to be able to reduce that problem, while it is not even one of the outcomes listed for the SOS program. If one of the desired outcomes of your intervention is to reduce or prevent suicidal thoughts and behaviors, then the Problem Solving Therapy program is shown to be effective, while the SOS program only has promising results.

Also before making a choice, you will want to read the program descriptions. This information can be accessed by looking in the column farthest to the right ("Program Description") on the "more info" link. You can access the same information by clicking on the hot-linked program title in the first column. Doing so takes you to a web page with a considerable amount of information on the program. (See Figure 6.4 for a portion of the information on the Problem Solving Therapy program.) You can read a complete program description, review evaluation findings by outcome, peruse the evaluation methodology, determine the references used, and learn about resources for program dissemination and implementation, including potential costs for training and materials. If you have many choices for potential programs that will positively address outcomes important to achieve in your community, you will want to look carefully at the information available within NREPP.

This process of refining results in NREPP is very quick. The example shown here went to a highly targeted list of just two programs within a few minutes. NREPP is, of course, a limited number of programs, but if you can find an appropriate EBP you can quickly develop this section of your grant proposal as well as have important information for the budget and other parts of your submission.

Remember that other government agencies have their own lists of programs that are considered evidence-based. Here are a few to examine on your own.

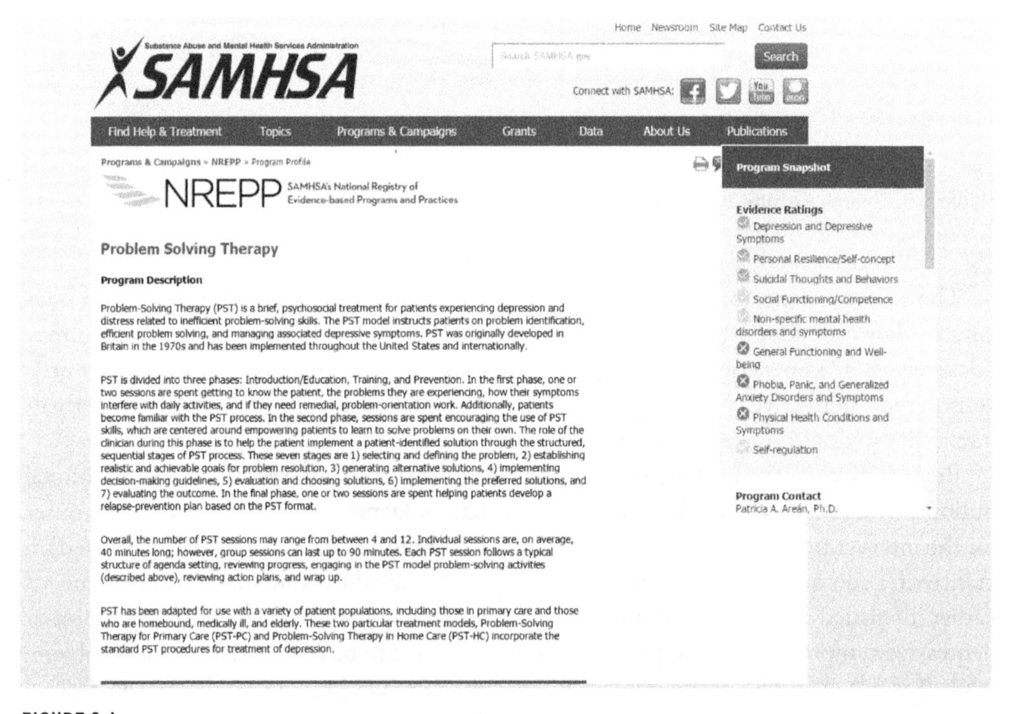

FIGURE 6.4: Problem Solving Therapy Program Information

CRIMESOLUTIONS

If you're writing a grant related to criminal justice, juvenile justice, or victims services topics, the US Department of Justice has its own database of effective programs. It's called www.CrimeSolutions.gov. It is "a central, reliable resource to help you understand what works in justice-related programs and practices."

MODEL PROGRAMS GUIDE

The office of Juvenile Justice and Delinquency Prevention provided descriptions of EBPs called the Model Programs Guide, available at http://www.ojjdp.gov/MPG. On its homepage, this website is said to provide "information about evidence-based juvenile justice and youth prevention, intervention, and reentry programs. It is a resource for practitioners and communities about what works, what is promising, and what does not work in juvenile justice, delinquency prevention, and child protection and safety."

THE OFFICE OF ADOLESCENT HEALTH: HEALTH AND HUMAN SERVICES TEEN PREGNANCY PREVENTION EVIDENCE REVIEW

The Office of Adolescent Health (OAH) provides a listing of programs with impacts on teen pregnancies or births, sexually transmitted infections (STIs), or sexual activity. Updated in April 2016, it is located at http://www.hhs.gov/ash/oah/oah-initiatives/teen_pregnancy/db/tpp-searchable.html.

ADMINISTRATION FOR CHILDREN AND FAMILIES: HOME VISITING, EVIDENCE OF EFFECTIVENESS

The Administration for Children and Families website reviews evidence of effectiveness for specific home-visiting program models. Currently, there are over 40 models with evidence to support their effectiveness—even when they say evidence is lacking. The information relating to program outcomes is especially interesting. Find out how the programs impact child development and school readiness; child health; family economic self-sufficiency; linkages and referrals; maternal health; positive parenting practices, reductions in child maltreatment; and reductions in juvenile delinquency, family violence, and crime. This is all available at http://homvee.acf.hhs.gov/outcomes.aspx.

THE NATIONAL COUNCIL ON AGING'S CENTER FOR HEALTHY AGING

The National Council on Aging is a nongovernmental agency and has only a few evidence-based programs in its listing. Still this is a promising start, and the website also has information on what EBPs are and why they are important. This website is at https://www.ncoa.org/center-for-healthy-aging/basics-of-evidence-based-programs/about-evidence-based-programs/.

SEXUALITY INFORMATION AND EDUCATION FOR TEENS

Another nongovernmental organization providing a listing of evidence-based programs in its area of interest is the Sexuality Information and Education Council of the United States (SIECUS). Thirty-five programs primarily relating to teens and sexuality are listed here. The website is http://siecus.org/index.cfm?fuseaction=page.viewPage&pageID=1484&nodeID=1.

THE CALIFORNIA EVIDENCE-BASED CLEARINGHOUSE FOR CHILD WELFARE

The mission of the California Evidence-Based Clearinghouse for Child Welfare is "to advance the effective implementation of evidence-based practices for children and families involved with the child welfare system." The visitor to its website (http://www.cebc4cw.org/) can search its database of programs ("view programs") as well as search for information to assist in selecting and implementing programs. It is funded by the California Department of Social Services' Office of Child Abuse Prevention.

THE CAMPBELL COLLABORATION

Another source of evidence-based programs and practices can be found on the website of the Campbell Collaboration (http://www.campbellcollaboration.org/) (see Figure 6.5). The organization's tagline on its website is: "What helps? What harms? Based on what evidence?" The mission of the Campbell Collaboration is to help people "make well-informed decisions by preparing, maintaining and disseminating systematic reviews in education, crime and justice, social welfare and international development" (Campbell Collaboration, n.d.).

While the contents of the page are interesting, the most important element is circled in Figure 6.5 near the top. This is the link to the Campbell Collaboration's library of systematic reviews of high-quality research to answer questions regarding the evidence base for solving various social problems. There is a search box lower on the page (see "X" in Figure 6.5) but this searches the entire website, not just the systematic reviews, which are

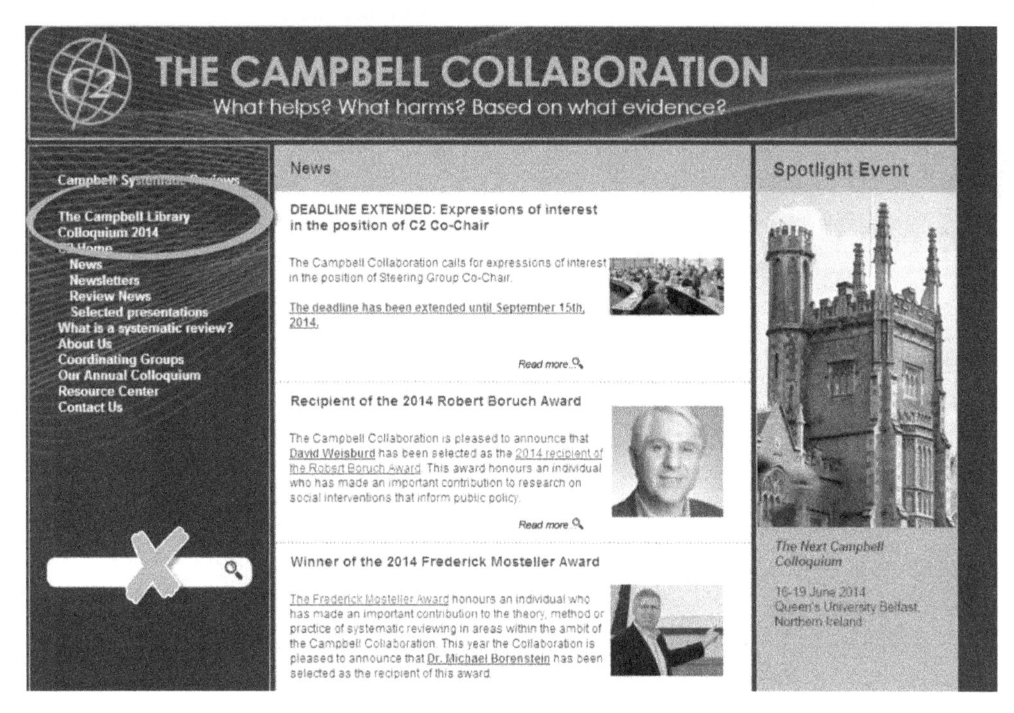

FIGURE 6.5: Campbell Collaboration Homepage

all we are interested in. Thus, it is easier to look for systematic reviews that have been conducted under the auspices of the Campbell Collaboration by first going to the Campbell Library link. When this is done, you can then search for a term in the library's database (see red circle in Figure 6.5). For this example, I use the term "self-esteem" (see results in Figure 6.6).

According to the second review study in our search results (Ekeland, Abbott, Hagen, Heian, & Nordheim, 2005):

> Some evidence exists that exercise has positive short-term effects on self-esteem in children and young people. Improving self-esteem may help to prevent the development of psychological and behavioral problems which are common in children and adolescents. Strong evidence exists for the benefits of exercise on physical health, but evidence for the effects of exercise on mental health is scarce. This review of trials suggests that exercise has positive short-term effects on self-esteem in children and young people, and concludes that exercise may be an important measure in improving children's self-esteem. However, the reviewers note that the trials included in the review were small-scale, and recognize the need for further well-designed research in this area. (Campbell Collaboration, n.d., Abstract, http://campbellcollaboration.org/lib/project/8/)

This type of information shows the strengths and limitations of the Campbell Collaboration quite strongly. The authors are top-notch, and the reviews are both deep and broad. All (or nearly all) relevant studies on the particular subject are found, analyzed,

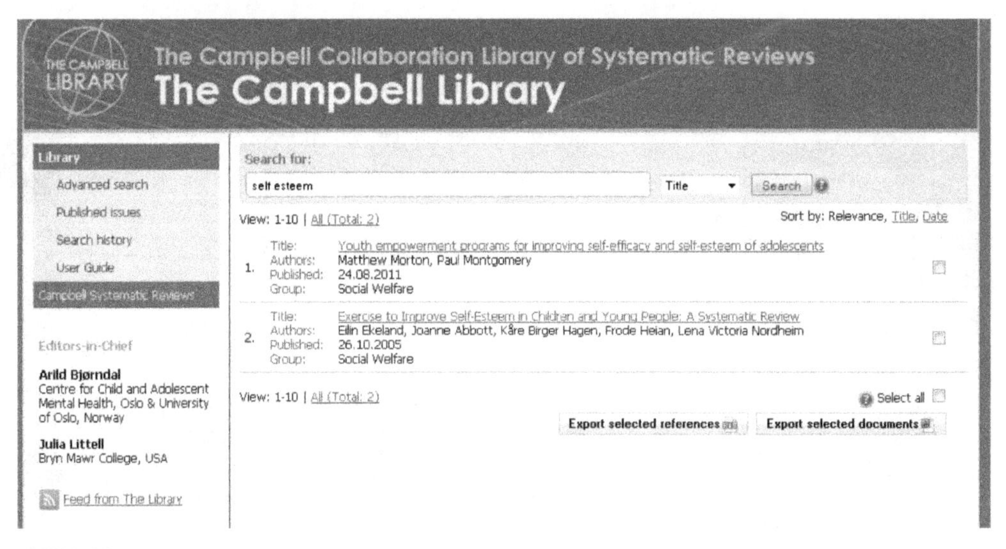

FIGURE 6.6: Results of Searching for "Self-Esteem" in the Campbell Collaboration

and discussed, with a clear conclusion being written, even if the conclusion is "We don't yet have enough good research to answer our question." As a source for results that are easy to find and apply, it is not as practical as NREPP, which has prepackaged solutions in the form of actual programs.

If you and/or your team are good at translating basic principles from a research setting to a programmatic setting, however, you can find valuable information for putting together an evidence-base for a new or adapted program using the Campbell Collaboration library. Let's use the conclusion regarding the use of exercise to increase self-esteem provided in the second systematic review in our search (Ekland et al., 2005), cited earlier as an example. If you wanted to write a grant to reduce childhood obesity in your community, this systematic review provides support for some secondary (and valuable) outcomes besides using exercise to lose weight. Not only would obese students lose weight, you could argue, but it is likely they will have an increase in self-esteem, which could have positive results in terms of grades in school, staying at grade level with chronological peers, graduating from high school, and job readiness. If you could find other supplementary research evidence to support these later desired outcomes, you may be able to propose and argue that a relatively small investment in physical activity now could have very large benefits later on based on credible research.

THE COCHRANE COLLABORATION

Another source of evidence for programs comes from the Cochrane Collaboration, which is similar to the Campbell Collaboration although with a different focus area. The Cochrane Collaboration focuses on health and mental health issues rather than social welfare or international development issues as the Campbell Collaboration does. Its website can be found at http://www.cochrane.org.

Without going into the detail of the previous sections (because the process is very similar), searching the Cochrane Collaboration, while not as easy as using the NREPP, is about as easy as searching the Campbell Collaboration. Looking through the literature reviews in the Cochrane Collaboration, I found one focused on the impact of exercise on depression in adults. Because the topic is similar to what we examined from the Campbell Collaboration, it is interesting to compare the findings.

The review is called "Exercise for Depression" by Rimer et al. (2012). The authors conclude:

> Exercise seems to improve depressive symptoms in people with a diagnosis of depression when compared with no treatment or control intervention; however since analyses of methodologically robust trials show a much smaller effect in favor of exercise, some caution is required in interpreting these results.

When combined with information on exercise in adolescents as seen in the review shown in the Campbell Collaboration, we see a possible set of findings in support of exercise for self-esteem and fighting depression at different ages. Thus, grantwriters can begin to develop an idea for a new program to address the issue of exercise and depression, one that is evidence-based but is not an EBP that has already been through a research process.

Using the Campbell and Cochrane Collaborations is not as easy as using NREPP, but may be just the approach needed if you cannot find a program within NREPP that addresses the needs you have found in your community. You'll have to translate some of the research findings into programmatic decisions and test them without the luxury of being able to follow someone else's footsteps exactly. But since many foundations and government grants want to showcase "innovation" and "new ideas" to solving social problems, this way of learning about the underlying research behind a program and then turning it into your own program ideas may just be the ticket to sizable grants.

Since the NREPP only includes substance abuse and mental health programs; the Campbell Collaboration emphasizes crime and justice, education, international development, and social welfare; and the Cochrane Collaboration focuses on medical and mental health problems, the question comes up frequently: Where shall I go when my grant is in another area? The next section discusses how to use search engines to discover other evidence-based program ideas.

SEARCH ENGINES

We all use search engines to find information on the Internet, but we may not have thought about what they are. A search engine is a software program that searches documents for specified keywords and returns a list of matches. Google is the most popular search engine in the world based on the number of queries it handles each day, but it is by no means the only search engine around. Some of the other commonly used search engines are Bing (www.Bing.com) and Yahoo (www.yahoo.com). For a seemingly comprehensive list of search engines, go to http://www.thesearchenginelist.com/. All search engines have their pros and cons. For the sake of comparison, we will look at the three major search engines' results for the same keywords (Figure 6.7, Figure 6.8, and Figure 6.9).

Google, although the most popular, has at least one interesting quirk in how it works that can be a problem. In an effort to return the most relevant results to the person doing the query, Google also searches through the person's other recent queries and the web pages that have been viewed on that computer to "match" the results to other recent searches. Google also does this with the ads it shows. Thus, there is a danger when using Google that it will exclude some important or valuable results if they are not related to your recent Google search history. For this reason, it is very important to use more than one search engine to look for results. To its credit, Google is constantly altering its search engine algorithms to try to return higher quality information to the user, thus prioritizing websites with a higher "authority" ranking.

Looking at Figure 6.7, the search results from Google, you can see that these are the first few results from a total of "about 11,200,000." We see there is an advertisement above the useful search results related in content to the search keywords. Underneath the advertisement is a short list of "scholarly articles" on the topic that come from Google's "Google Scholar" search engine. This is a tool that can help you avoid sifting through so many non-evidence-based results.

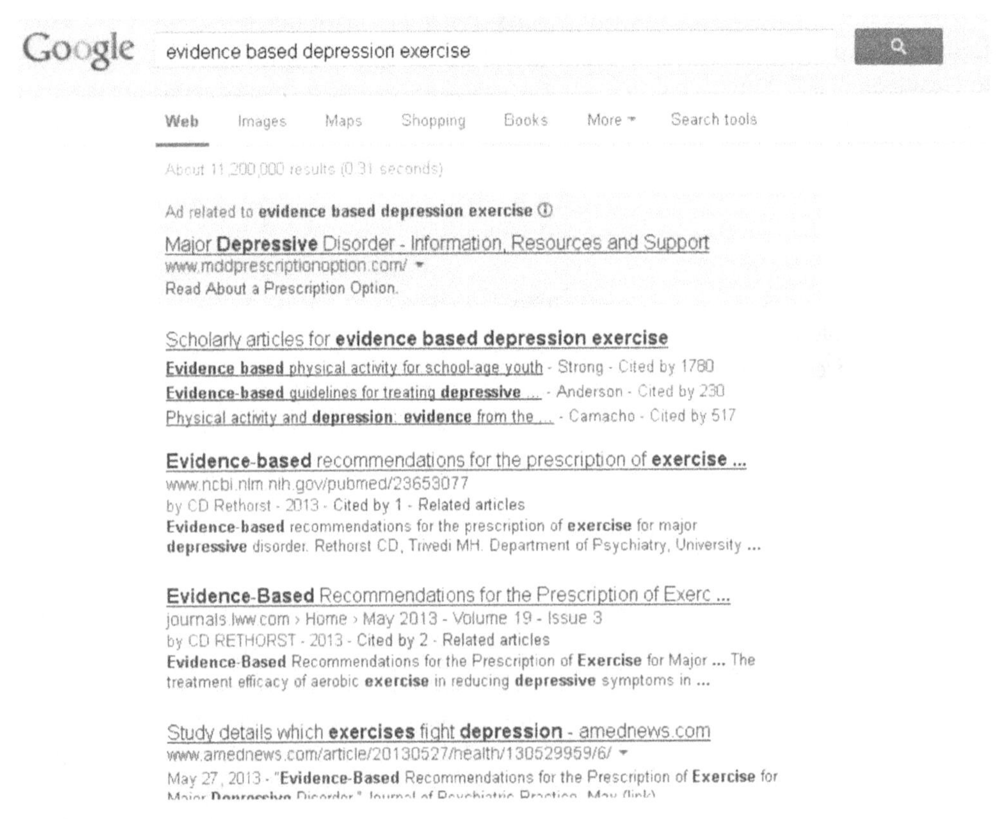

FIGURE 6.7: Google Search Engine Results

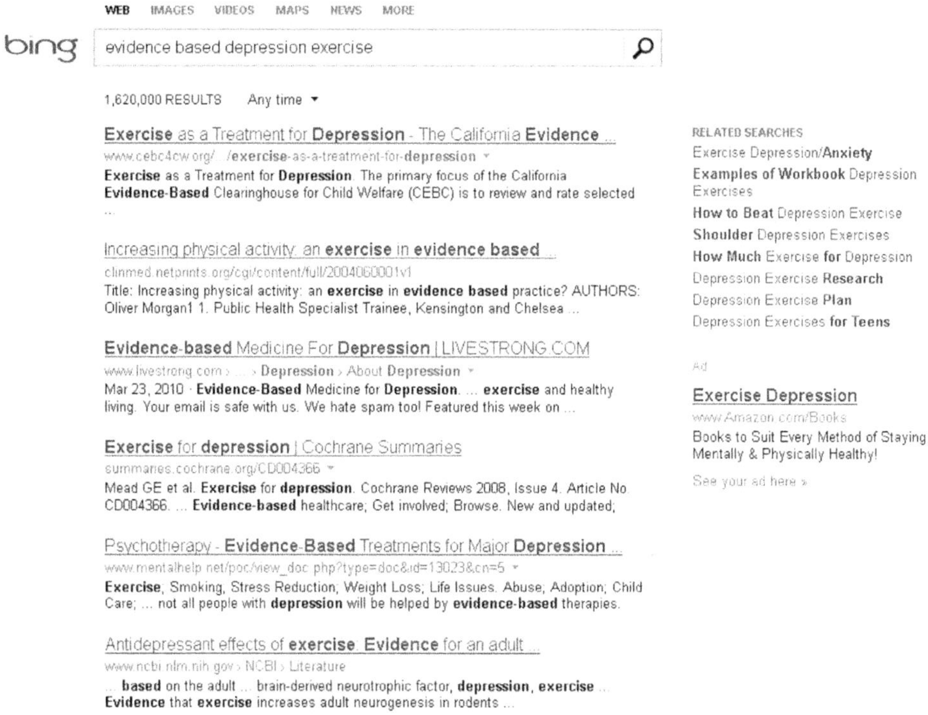

WEB IMAGES VIDEOS MAPS NEWS MORE

bing evidence based depression exercise

1,620,000 RESULTS Any time ▾

Exercise as a Treatment for **Depression** - The California **Evidence** ...
www.cebc4cw.org/.../**exercise**-as-a-treatment-for-**depression** ▾
Exercise as a Treatment for **Depression**. The primary focus of the California **Evidence-Based** Clearinghouse for Child Welfare (CEBC) is to review and rate selected ...

Increasing physical activity: an **exercise** in **evidence based** ...
clinmed.netprints.org/cgi/content/full/2004060001v1
Title: Increasing physical activity: an **exercise** in **evidence based** practice? AUTHORS: Oliver Morgan1 1. Public Health Specialist Trainee, Kensington and Chelsea ...

Evidence-based Medicine For **Depression** | LIVESTRONG.COM
www.livestrong.com › ... › **Depression** › About **Depression** ▾
Mar 23, 2010 · **Evidence-Based** Medicine for **Depression**. ... **exercise** and healthy living. Your email is safe with us. We hate spam too! Featured this week on ...

Exercise for **depression** | Cochrane Summaries
summaries.cochrane.org/CD004366 ▾
Mead GE et al. **Exercise** for **depression**. Cochrane Reviews 2008, Issue 4. Article No. CD004366. ... **Evidence-based** healthcare; Get involved; Browse. New and updated;

Psychotherapy - **Evidence-Based** Treatments for Major **Depression** ...
www.mentalhelp.net/poc/view_doc.php?type=doc&id=13023&cn=5 ▾
Exercise; Smoking; Stress Reduction; Weight Loss; Life Issues. Abuse; Adoption; Child Care; ... not all people with **depression** will be helped by **evidence-based** therapies.

Antidepressant effects of **exercise**: **Evidence** for an adult ...
www.ncbi.nlm.nih.gov › NCBI › Literature
... **based** on the adult ... brain-derived neurotrophic factor, **depression**, **exercise** ... **Evidence** that **exercise** increases adult neurogenesis in rodents ...

RELATED SEARCHES
Exercise Depression/**Anxiety**
Examples of Workbook Depression Exercises
How to Beat Depression Exercise
Shoulder Depression Exercises
How Much Exercise **for** Depression
Depression Exercise **Research**
Depression Exercise **Plan**
Depression Exercises **for Teens**

Ad
Exercise Depression
www.Amazon.com/Books
Books to Suit Every Method of Staying Mentally & Physically Healthy!
See your ad here »

FIGURE 6.8: Bing Search Engine Results

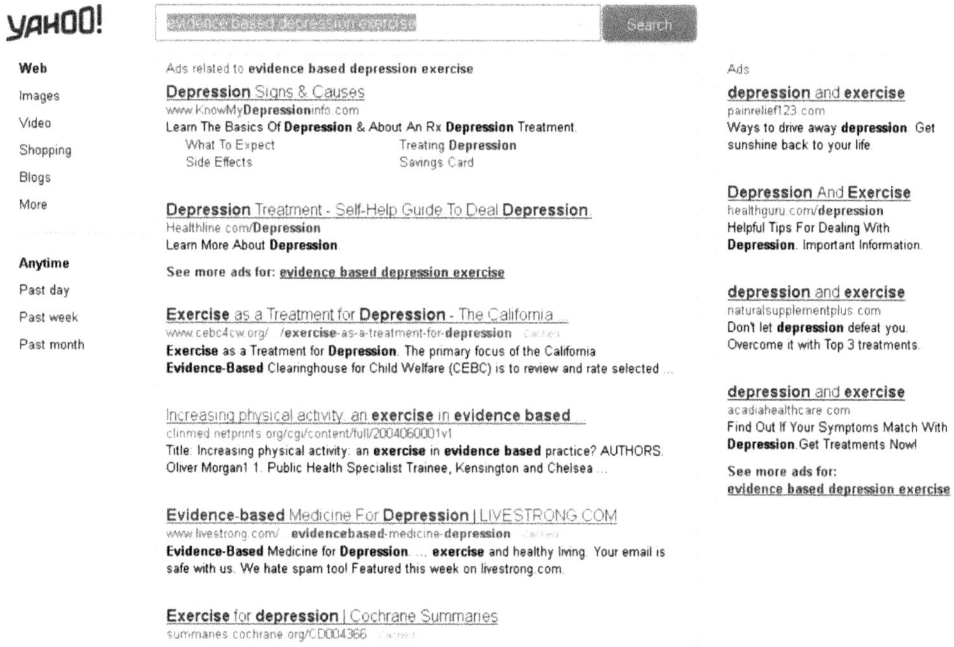

YAHOO! Search

Web
Images
Video
Shopping
Blogs
More

Anytime
Past day
Past week
Past month

Ads related to **evidence based depression exercise**

Depression Signs & Causes
www.KnowMy**Depression**info.com
Learn The Basics Of **Depression** & About An Rx **Depression** Treatment.
What To Expect | Treating **Depression**
Side Effects | Savings Card

Depression Treatment - Self-Help Guide To Deal **Depression**.
Healthline.com/**Depression**
Learn More About **Depression**.

See more ads for: **evidence based depression exercise**

Exercise as a Treatment for **Depression** - The California ...
www.cebc4cw.org/.../**exercise**-as-a-treatment-for-**depression**
Exercise as a Treatment for **Depression**. The primary focus of the California **Evidence-Based** Clearinghouse for Child Welfare (CEBC) is to review and rate selected ...

Increasing physical activity: an **exercise** in **evidence based** ...
clinmed.netprints.org/cgi/content/full/2004060001v1
Title: Increasing physical activity: an **exercise** in **evidence based** practice? AUTHORS: Oliver Morgan1 1. Public Health Specialist Trainee, Kensington and Chelsea ...

Evidence-based Medicine For **Depression** | LIVESTRONG.COM
www.livestrong.com/... **evidencebased**-medicine-**depression**
Evidence-Based Medicine for **Depression**. ... **exercise** and healthy living. Your email is safe with us. We hate spam too! Featured this week on livestrong.com.

Exercise for **depression** | Cochrane Summaries
summaries.cochrane.org/CD004366

Ads
depression and **exercise**
painrelief123.com
Ways to drive away **depression**. Get sunshine back to your life.

Depression And **Exercise**
healthguru.com/**depression**
Helpful Tips For Dealing With **Depression**. Important Information.

depression and **exercise**
naturalsupplementplus.com
Don't let **depression** defeat you. Overcome it with Top 3 treatments.

depression and **exercise**
acadiahealthcare.com
Find Out If Your Symptoms Match With **Depression**. Get Treatments Now!

See more ads for:
evidence based depression exercise

FIGURE 6.9: Yahoo Search Engine Results

The first result that comes up is a summary article from a very authoritative source, the National Institutes of Health. Interestingly, the second result is another reference to the same source, while the third result is yet another way to get to the same basic information, but via a press release to the journal article. With the Google search engine, you can quickly search for YouTube videos on the subject by clicking on "YouTube" above the search box without retyping the search terms. In this case, 26,400 video results are found regarding "evidence-based depression exercise" which may be a useful adjunct to learning information for your grant proposal.

Looking next at Figure 6.8, the search results from Bing, we see about 5,000,000 more results returned than from Google. Now, it isn't as if we are going to search all 16.2 million results anyway, but it is interesting to see this vastly different number. Bing shows "related searches" along the right and also has the ability to click on a search of videos easily.

The first result returned (using the same keywords in the search) is the site of the California Evidence-Based Clearinghouse for Child Welfare, which doesn't appear in the first few results of Google at all. The second result is to an unpublished manuscript of an article on community medicine, while the third is a link to a Livestrong.com page discussing whether depression medications can cause anger issues, which is not truly on-target for the subject we are looking for. Bing's fourth result is a link to the Cochrane Collaboration's review discussed earlier in this chapter. It is not until result number six in Bing that we see Google's first result.

Turning finally to Yahoo's results (Figure 6.9), we see that there are advertisements along the right side and above the search results, as with Google. The results listed are similar to the ones seen in Bing. This is because Yahoo's search engine is now "powered by Bing," a fact noted at the bottom of the first page of results (not shown). Yahoo, however, seems to have more advertisements than Bing.

Comparing these three search engines shows that you will come across different results using different products. Metasearch engines have been developed that do the searches on the different platforms and then aggregate the results for you. One example of this is Dogpile (www.dogpile.com) which "fetches" results for the user from Google, Yahoo, and Yandex, and possibly other, more specialized search engines (see Figure 6.10). The advantages of a metasearch engine are obvious: It offers "one-stop shopping." The key is to be aware of such options and to be willing to try them. In the end, the choice of one or more search engines to find information on EBPs is going to be based on individual preference.

OTHER CONSIDERATIONS WHEN FINDING AND CREATING EVIDENCE-BASED PROGRAMS

So far in this chapter, we have explored using listings of existing programs with support from research (NREPP, for example), exploring sites featuring systematic reviews of research relating to specific problems (Campbell Collaboration and Cochrane Collaboration) and examining results found from using search engines. These are all

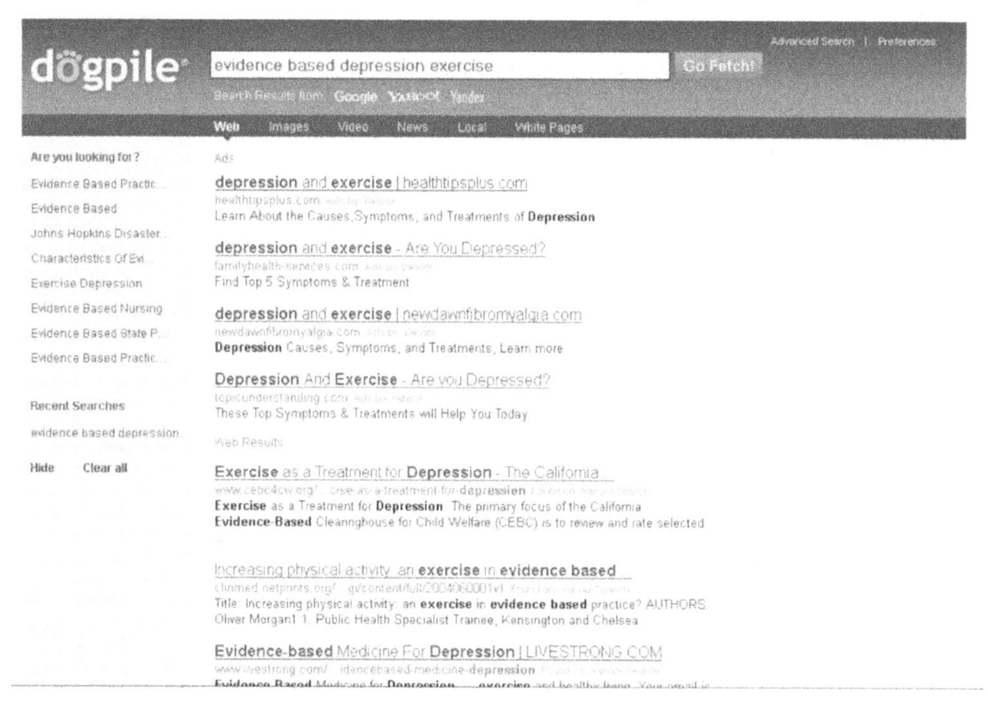

FIGURE 6.10: Dogpile Metasearch Engine Search Results

fairly systematic approaches that can yield good results as you think about how to address problems in your community.

There is another approach that is less systematic but highly recommended. You probably already read the professional and academic literature for ideas and information about similar programs. But you should also be on the lookout for information on promising program ideas in the newspaper and in magazines you can pick up at the newsstand. These latter outlets frequently want to showcase positive human interest stories and often describe a program that helped a particular client achieve success in overcoming problems. When you read or view a story that pertains to social issues similar to yours, be sure to note the program name and follow up with a search of the organization running it. You can gather additional information from its website and even e-mail or phone to find out more.

It is very useful to keep a scrapbook or idea file where you can place summaries of programs that you discover in both professional and popular media. In your summaries, start with where you heard about the program so you can track it down again when you need to. Write in your summary what the goal of the program is, what program components you can uncover, and any other information that you have been able to find, including the organization running the program and a contact name if you have located one. If you file these in some nonrandom way corresponding to how they connect with your organization's priorities and goals, you will have a good starting point for generating additional ideas when it comes time to writing a future grant.

Before you definitively choose a program to include in your grant proposal, you will want to determine if it is a fit for your organization and community in other ways, as well. The NREPP ("Questions to Ask: CSAP's Checklist for Program Fit") mentions 10 additional things to consider when you are in the process of selecting a program:

- Your population
- Your setting
- Your target population's culture
- Availability of program manuals, guides, training, and evaluation tools
- Values of the community
- Values of the local power structure
- Organizational mission, vision, and culture
- Administrative feasibility
- Capacity of staff
- Availability of sufficient financial, time, and other resources.

The Family and Youth Services Bureau (n.d.) also recommends a number of steps to take prior to selecting your program to ensure that the program fits in well with your organization and community.

SUMMARY

When writing the "solution" section of your grant proposal, it is much easier to find a predesigned program with evidence to support its efficacy. These can be hard to discover however, when you have specific goals for a clearly defined subpopulation. Adapting an EBP can get around some of the problems, but you must be careful not to change core components of the program or else you undermine the reason for choosing an EBP.

The creation of an idea for a program based on the research you've done can be both exhilarating and frustrating. When you uncover and apply both long-established information and the latest research findings, when you are able to link ideas previously located in separate silos, you play the role of explorer, discoverer, and assimilator. Taking research-based information and molding it into a set of ideas for staff to implement and clients to use, while maintaining community and funder approval is definitely a challenge. There will be gaps in your knowledge and the research base no matter how diligently you strive to fill them. You'll have to work inside of a large zone of ambiguity and doubt. You'll have to be persuasive in defending your nascent ideas even while being open to constructive criticism that makes your concepts clearer and more likely to both receive start-up resources and succeed once implemented. These are challenging but essential tasks in the job of a grantwriter.

PRACTICE WHAT YOU'VE LEARNED

1. Conduct a search of one or more established listings of evidence-based practices and programs (such as NREPP) for a program that addresses some aspect of the problem you wrote about when practicing what you learned in the previous chapter. If you don't find any appropriate programs in your first search, find at least one other registry and try there. (If you can't find anything remotely appropriate, write down what you've tried and then move on to Exercise 2.) Search until you have found at least two programs that seem promising. When you have done this, copy the information about them and gather additional information on them. Make a table where you compare the programs in terms of their goals, components, population, and so on, as described in this chapter.
2. Using the Cochrane Collaboration and the Campbell Collaboration libraries, as well as other depositories and academic journals, look for systematic research summaries and journal articles that address topics relevant to your community's problems. Find at least five sources of evidence-based ideas for solving the problem. Write down the source of the information, the key ideas and research findings, and how these findings can be used in a program to address the problems of your community.
3. Finally, using one or more search engines, find additional resources to put together a program that addresses the problem you wish to solve. Make a comprehensive list of your materials including source, summary or abstract, and the names of any organizations you think could provide additional information.

REFERENCES

Campbell Collaboration (n.d.). About Us. http://www.campbellcollaboration.org/about_us/index.php

Cochrane Collaboration (n.d.). Home. http://www.cochrane.org

Ekland, E., Abbot, J., Hagen, K., Heian, F., & Nordheim, L. (2005). Exercise to Improve Self-Esteem in Children and Young People: A Systematic Review. Retrieved from http://campbellcollaboration.org/lib/project/8/

Family and Youth Services Bureau (n.d.). Selecting an Evidence-Based Program That Fits Tip Sheet. Retrieved from http://www.acf.hhs.gov/sites/default/files/fysb/prep-program-fit-ts.pdf

Gambrill, E. (1990). *Critical Thinking in Clinical Practice: Improving the Accuracy of Judgments and Decisions About Clients*. San Francisco: Jossey-Bass.

Gambrill, E. (1999). Evidence-based practice: An alternative to authority-based practice. *Families in Society, 80,* 341–350.

Gibbs, L. & Gambrill, E. (2002). Evidence-based practice: Counterarguments to objections. *Research on Social Work Practice, 12,* 452–476.

Hoefer, R. & Jordan, C. (2008). Missing links in evidence-based practice for macro social work. *Journal of Evidence-based Social Work, 5,* 549–568.

National Registry of Evidence-Based Programs and Practices (NREPP) (2013). www.nrepp.samhsa.gov

Office of Justice Programs (2014). Youth with Sexual Behavior Problems. OJJDP FY 2014–3818. https://www.ojjdp.gov/grants/solicitations/FY2014/YSBP.pdf

Rimer, J., Dwan, K., Lawlor, D. A., Greig, C. A., McMurdo, M., Morley, W., & Mead, G. E. (2012). Exercise for depression. *Cochrane Database of Systematic Reviews,* 7. Art. No.: CD004366. DOI: 10.1002/14651858.CD004366.pub5 –

Sackett, D., Richardson, W., Rosenberg, W., & Haynes R. (1997). *Evidence-Based Medicine: How to Practice and Teach EBM*. New York: Churchill Livingstone.

Substance Abuse and Mental Health Services Administration (SAMHSA) (n.d.). What Are Core Components and Why Do They Matter? Retrieved from http://captus.samhsa.gov/access-resources/what-are-core-components-and-why-do-they-matter

CHAPTER 7

LOGIC MODELS

Now that you have a program that you want to write a grant to implement, you want to be able to quickly and easily communicate the connections between the need you've uncovered, the resources you're asking for, what you will do with the funds you receive, and the changes you expect will occur for clients because of the program. You could write a narrative about all these things (and you will) but there is an easier way to describe the program than with all those words. It's called a "logic model," and it is a quickly understood graphical representation of your program, connecting the problem to the solution you propose. It does not take the place of a narrative description in your grant proposal, but it does make understanding your program a lot easier for potential reviewers and funders who will decide whether you receive the resources you are requesting.

The idea of logic models as an adjunct to program development extends at least as far back as 2000, when the Kellogg Foundation published a guide to developing logic models for program design and evaluation, although the newest version was updated in 2004 (Kellogg Foundation, 2004). While you can find many variations describing how a logic model should be constructed, it is a versatile tool that is used to design programs, assist in their implementation, and guide their evaluation. This chapter describes one basic approach to logic modeling. We examine its use for program design in this chapter and, in later chapters, we use it to understand program implementation and evaluation tasks.

At the onset, you should understand that not all programs have been designed with the aid of a logic model, although the practice is becoming more common every year. Federal grants, for example, usually require applicants to submit a logic model, and their use throughout the human services sector is growing because of education and in-service training. If there is no logic model for a program you are working with, it is possible to create one *after* a program has been implemented. In this way, you can use an evidence-based program model even if it does not "come with" a logic model as part of the package. If you are designing a program from the ground up, this is certainly an essential technique.

LOGIC MODELS AND THEORIES OF CHANGE

According to Frechtling (2007), a logic model is "a tool that describes the *theory of change* underlying an intervention, product or policy" (p. 1, emphasis in the original). A theory of change is the set of processes you believe will happen so that a program will work to improve clients' lives. Because logic models are said to describe the program's "theory of change," it is possible to believe that this refers to a big popular theory like social learning theory, cognitive-behavioral theory, or any one of a number of psychological, political, or sociological theories. While the logic of the proposed program is likely to be based to some extent on a grand theoretical perspective such as these, the theory of change as shown in a logic model is much less expansive.

Let's take a simple example of a theory of change: Teenage smoking is a problem. In order to decrease the problem, an intervention could be for adults to tell the teens not to smoke. The logic of this is clear: teens that are told to stop doing things by adults actually stop doing those things. Do you see any problems with this as a viable "theory of change?" We could make the intervention more complicated by adding additional strategies, such as passing stricter laws against the sale of tobacco to teens and young adults or by requiring anti-tobacco messages in schools. Each additional strategy is another hypothesis about what will decrease the problem of teen smoking. They are each grounded in a "theory of change" that may or may not be correct.

A logic model is a visual representation of all the hypotheses that are implicit in a program. For example, a program may operate on the theory that receiving education about using condoms to prevent the spread of HIV/AIDS will be enough to ensure their use, but this is really just a hypothesis. People receiving a certain educational experience ***may*** or ***may not*** act in accordance with the education they received. From the program designer's viewpoint, there is a strong likelihood that people will act in accord with their education. However, until the program is put into place and data are gathered about actual behavior, the connection is just a theorized hypothesis.

The word "theory" implies it is not *certain* that everything will work as planned. As a result, using a logic model to discuss how the program designer thinks things will occur is important. Using this process, gaps in logic can come to the forefront before the program begins, instead of after. People may know of research or have experiences that contradict the predicted connection, in which case there may be a need to change the design of the program before it begins. In addition to gaps or errors in the logic of the proposed program, additional things can occur, which are called "unanticipated outcomes." These are changes to clients that were not considered when the program was planned. Unanticipated outcomes may be positive or negative.

Some unanticipated outcomes almost always occur, but it is better to keep them to a minimum, particularly negative unanticipated outcomes. One way to do this is to talk to potential clients and staff members with experience working with the client population about what clients might experience going through the program. As a grantwriter it may be easy to become isolated from the day-to-day aspects of life on the "front lines" so it is beneficial to gather these perspectives as you develop the logic model. If you are able to identify additional positive outcomes, these can be placed in the logic model; thus, you can better measure and understand how your program works and what it achieves. If you identify initially unanticipated potential negative outcomes, you can work on ways to prevent them from happening.

WHAT IS A LOGIC MODEL?

According to the Office of Juvenile Justice and Delinquency Programs (OJJDP, 2014), every applicant must provide a logic model that "graphically illustrates how the performance measures are related to the project's problems, goals, objectives, and design" (p. 20). Another federal agency states the following in a request for proposals:

> Applicants must submit a logic model for designing and managing their project. A logic model is a tool that presents the conceptual framework for a proposed project and explains the linkages among program elements. While there are many versions of the logic model, they generally summarize the logical connections among the needs that are the focus of the project, project goals and objectives, the target population, project inputs (resources), the proposed activities/processes directed toward the target population, the expected short- and long-term outcomes the initiative is designed to achieve, and the evaluation plan for measuring the extent to which proposed processes and outcomes actually occur. (Administration on Children, Youth and Families, 2014, p. 21)

The logic model you will be able to develop by the end of this chapter will look much like Figure 7.1. It begins with a statement of the problem or need to inform the reader why the program has been developed. From there, logic models show how you propose to use the resources you have to make the changes in the clients that you believe are needed.

The utility of a logic model is to show how the resources used (inputs) are changed into a program (activities) with closely linked products (outputs) that then lead to changes in clients in the short, medium, and long terms (outcomes). The net effect of these client changes is that the original problem is solved or at least made better for the clients in the program. An example logic model is shown as Figure 7.1 and explained in what follows.

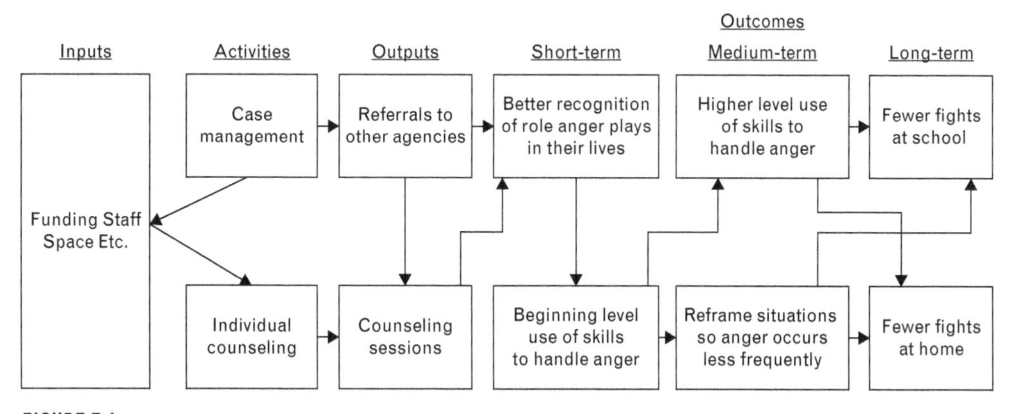

FIGURE 7.1: Example Logic Model

THE PROBLEM STATEMENT

We begin by placing the need for the program (the problem) on the line labeled "Problem" so that, in the end, we can show how we intend to solve (or at least make better) the key issue affecting the population we have chosen to work with. The problem should be clearly stated in a way that does not predetermine how the problem will be solved. For example, if you write that the problem you are dealing with is "A lack of [something] for [some population group]" you will not have a strong problem statement that could be solved in a number of ways. This formulation of the problem leads to only one solution: Get more of that [something] and provide it to [some population group].

Looking at the example logic model's problem statement, we see it reads, "School-aged youth do not properly manage their anger, which leads to verbal and physical fights at school and home." This problem statement focuses on what the behaviors are that are a problem (verbal and physical fights), is specific about who has the problem (school-aged youth), what the initial cause of the problem is (inability to properly manage anger), and where it is a problem (school and home). It also does not prejudge what the solution is, and thus allows for many possible different programs to address the problem, each using different techniques and approaches.

Here is an example of a poorly worded problem statement: "There is a lack of anger management classes in schools for school-aged youth." It is not as useful as the first problem statement in the previous paragraph because it states the problem in a way that allows only one solution—get more anger management classes.

Another way to enhance the problem statement is to phrase the statement in such a way that almost anyone can agree that it is actually a problem. The example problem statement might make this point more clearly by saying, "There are too many verbal and physical fights at school and home among school-aged youth." Phrased this way, there would be little doubt that this is a problem, even though the statement is not specific about the number of such fights or the causes of the fights. By not including the phrase "inability to properly manage anger" in the problem statement, the program designer can more readily come up with innovative ideas for how to solve the problem.

This way of stating the problem might lead to a host of other issues being addressed (instead of anger management issues) that might be leading to fights, such as overcrowding in the halls, gang membership, conflict over curfews at home, or anything else that might conceivably cause youth to fight at school or home. Be prepared to revise your first effort at the problem statement and seek input from interested stakeholders to be sure that you are tackling what is really considered the reason for the program. The problem statement is vital to the rest of the logic model and evaluation, so take the time to make several drafts to get full agreement.

After the problem statement, the logic model has six columns. Arrows connect what is written in one column to something else in the next column to the right or even within the same column. These arrows are the "logic" of the program. If the column to the left is achieved, then we believe that the element at the end of the arrow will be achieved. Each arrow can be considered to show a hypothesis that the two elements are linked. (The example presented here is intentionally not "perfect," so that you can see some of the nuances of using this tool.)

COLUMN 1: INPUTS

The first column is labeled "Inputs." In this column, you write the major resources that will be needed or used in the program. Generically, these tend to be funds, staff, and space, but they can also include other elements such as type of funds, educational level of the staff, and location of the space (on a bus line, for example), if they apply to your program. The resource of "staff," for example, might mean MSW-level licensed counselors. In the end, if only staff members with bachelor's degrees in psychology are hired, this would indicate that the "staff" input was inadequate.

COLUMN 2: ACTIVITIES

The second column is "Activities." In this area, you write what the staff members of the program will be doing—what behaviors you would see them engage in if you sat and watched them. Here, as elsewhere in the logic model, there are decisions about the level of detail to include. It would be too detailed, for example, to have the following bullet points for the case management activity:

- Answer phone calls about clients
- Make phone calls about clients
- Learn about other agencies' services
- Write out referral forms for clients to other agencies

This *is* what you would see, literally, but the phrase "case management" is probably enough. Somewhere in program documents there should be a more detailed description of the duties of a case manager, so that this level of detail is not necessary on the logic model, which is, after all, a graphical depiction of the program's theory of change, not a daily to-do list.

The other danger in listing activities on the logic model is being too general. In this case, a phrase such as "provide social work services" wouldn't be enough to help the reader know what the employee is doing because there are so many activities that are social work services. Getting the correct level of specificity is important in helping develop your implementation and evaluation plans after you create your logic model.

As you can see from the arrows leading from the inputs to the activities, the program theory indicates that, given the proper funds, staff, and space, the activities of case management and individual counseling will occur. This may or may not happen however, which is why a process evaluation is needed. (This topic is discussed in the chapter about evaluation.)

COLUMN 3: OUTPUTS

The third column in the logic model lists "Outputs." An output is a measurable result of an activity that is NOT a change in the client's knowledge, attitude or belief, status or behavior. In this example, the activity of "case management" results in client youth receiving "referrals to other agencies for services." The output of the activity "individual counseling" is "counseling sessions." It is important to note that outputs are not changes in clients—outputs are the results of agency activities that may or may not then result in changes to clients. Outputs can be counted and frequently are reported such as "In the past year, 150 clients were referred to a variety of youth services" or "Counselors provided 760 counseling sessions to clients." The connection between agency activity and outputs is perhaps the most difficult part of putting together a logic model because many people mistakenly assume that if a service is given and documented, then client changes are automatic. This is simply not true.

COLUMNS 4–6: OUTCOMES

The next three columns are collectively known as "Outcomes." An outcome is a change in the client and should be written in a way that is a change in knowledge, attitude, belief, status, or behavior. Outcomes are why programs are developed and run—to change clients' lives (see Box 7.1). Outcomes can be developed at any level of intervention—individual, couple or family, group, organization, or community of any size. This example uses a program designed to make a change at an individual youth level, but could also include changes at the school or district level if desired.

Outcomes are usually written to show a time dimension with short-, medium-, and long-term outcomes. The long-term outcome is the opposite of the problem stated at the top of the logic model and thus ties the entire intervention back to its purpose—to solve a particular problem. The division of outcomes into three distinct time periods is obviously a helpful fiction, not a tight description of reality. Still, some outcomes are expected to come sooner than others. These short-term outcomes are usually considered the direct result of outputs being developed.

In the example logic model, the arrows indicate that referrals and individual counseling are both supposed to result in client youth better recognizing the role that anger plays in their life. After that is achieved, the program theory hypothesizes that clients will use skills at a beginning level to handle their anger. This is a case where one short-term outcome (change in self-knowledge) leads the way for a change in behavior (using skills).

The element "beginning level use of skills to handle anger" has two arrows leading to medium-term outcomes. The first arrow leads to "higher level use of skills to handle anger." In this theory of change, at this point, there is still anger, but youth in the program recognize what is occurring and take measures to handle it in a skillful way that does not lead to negative consequences. The second arrow from "beginning level use of skills to handle anger" indicates that the program designers believe that the skills youth learn will assist them to reframe situations they are in so that they feel angry less frequently. This is a separate behavior than applying skills to handle anger, so it receives its own arrow and box.

The final column represents the long-term outcomes. Often, there is only one element shown in this column, which indicates the opposite of the problem. In this logic model, since the problem is seen to occur both at school and at home, each is looked at separately. A youth may reduce fights at home but not at school, or vice versa, so it is important to leave open the possibility of only partial success.

This example logic model shows a relatively simple program theory, with two separate tracks for intervention but with overlapping outcomes expected from the two intervention

BOX 7.1: OUTCOMES VERSUS GOALS AND OBJECTIVES: WHAT'S THE DIFFERENCE?

Logic models use the term "outcomes," but many people use the terms "goals" and "objectives" to talk about what a program is trying to achieve. You may have learned that an outcome objective answers the question, "What difference did it make in the lives of the people served?" In this chapter, you are told that an outcome is a "change in the client." What's the difference between "outcomes" and "goals and objectives?"

In reality, there is not much difference. Goals and objectives are one way of talking about the purpose of a program. This terminology is older than the logic model terminology and more widespread. But it can be confusing, too, because an objective at one level of an organization may be considered a goal at another level or at a different time.

Outcomes are easier to fit into the logic model approach to showing program theory by relating to resources, activities, and outputs. Systems theory terminology is more widespread now than before and avoids some of the conceptual pitfalls of goals and objectives thinking.

You will run into people and funders who prefer the terms "goals" and "objectives" and others who like "outcomes" as their term of choice. But you should realize that both approaches are ultimately talking about the same thing—the ability of an organization to make people's lives different and better.

methods. It indicates how one element can lead to more than one "next step" and how different elements can lead to the same outcome. Finally, while it is not necessarily obvious just yet, this example shows some weak points in the program's logic that will emerge when we use it as a guide to evaluating the program.

SUMMARY

Logic models are an important way to show a reader that your program makes sense. You help the funder see how your organization can take funding and turn it into improvements in clients' lives. All the steps in the theory of change are laid out in a way that invites the audience to step into the program, see how it would work, and begin to believe in its effectiveness, even before a dollar has been spent. It is not enough to show that a problem exists if you want to be funded. You must also demonstrate that you have a workable solution in mind. A well-developed logic model does this and takes you a big step along the way to having your proposal funded.

PRACTICE WHAT YOU'VE LEARNED

1. If you can, locate a logic model in a previous grant application or in the literature. Examine it carefully to see how it is similar to or different from the logic model approach that is described in this chapter. What elements are the same and which are different? What approach do you like better? Why?
2. Examine a program that you are familiar with. This might be where you currently work, have an internship, or used to work. The more familiar you are with it the better. Develop a logic model for it, following the process discussed in this chapter. Remember that it is usual to have to revise your problem statement and other aspects of your logic model more than once. When you have completed this task, share your logic model with someone else. Encourage your partner to ask tough questions that will force you to understand and explain the connections between one box or column and the next. Trade roles with your partner.
3. Locate a description of an evidence-based program that you may be considering for your grant. (Review finding evidence-based programs in chapter 6.) Develop a full logic model that has a problem statement and shows all the components of the program. Look at your logic model carefully, assessing whether any gaps in the logic appear. Explain your logic model to another person and request feedback and questions about the program.

REFERENCES

Administration on Children, Youth and Families (2014). Grants to Address Trafficking Within the Child Welfare Population. HHS-2014-ACF-ACYF-CA-0831. Retrieved from http://www.acf.hhs.gov/grants/open/foa/files/HHS-2014-ACF-ACYF-CA-0831_0.pdf

Frechtling, J. (2007). *Logic Modeling Methods in Program Evaluation.* San Francisco: Jossey-Bass.

Kellogg Foundation (2004). Logic Model Development Guide. Retrieved from http://www.wkkf.org/resource-directory/resource/2006/02/wk-kellogg-foundation-logic-model-development-guide

Office of Juvenile Justice and Delinquency Prevention (OJJDP) (2014). Youth with Sexual Behavior Problems Program (OJJDP-2014-3818). Retrieved from http://www.grants.gov/web/grants/search-grants.html

CHAPTER 8

PROGRAM EVALUATION

WHAT IS EVALUATION?

Evaluation is a way to determine the worth or value of a program (Rossi, Lipsey, & Freeman, 2003). In the Age of Scarcity we are experiencing, evaluation has become increasingly used as a means of ensuring accountability for spending government and foundation funds. Evaluation results can be used to show whether desired program outcomes are being achieved as well as for comparing the level of outcomes achieved by different programs. Well-designed and implemented evaluations improve individual programs and the mix of programs funded.

TYPES OF EVALUATION

The world of evaluation has progressed and continues to move forward, leaving in its wake terms and ideas about different types of evaluation. This book is not focused on the history of evaluation and so we won't go into detail, but you may run into a number of terms and phrases that relate to evaluation planning in grant proposals that differ from the terms used here. As always, it is most important to use the grant proposal instructions from the funder you are applying to. This chapter covers the three types of evaluations that many, if not most, grants will ask for: process, fidelity or implementation, and outcome.

PROCESS EVALUATION

One description of process evaluation tells us that it "looks at how program activities are delivered" (Substance Abuse and Mental Health Services Administration [SAMHSA], n.d.). Process evaluation ensures that you will be able to describe what and how many services were delivered to which clients. It may seem unnecessary, at first glance, because results are more important, but without knowing how the program was operated, it is difficult to say the program caused whatever changes in clients are seen. If client changes are not noted, it may be because the program itself was not put into effect properly, or with an adequate level of resources. Or it may be that the program theory shown in its logic model is incorrect. A process evaluation can help you tell which of these cases is correct. The term "implementation evaluation" is sometimes used as a synonym for process evaluation, as is the term "formative evaluation."

Process evaluations focus on the connections between "inputs," "activities," and "outputs" of the logic model. These are the columns on the left of the logic model.

FIDELITY ASSESSMENT

Fidelity assessments are, in some ways, a subset of process evaluation or implementation evaluation. They are frequently connected specifically to the implementation of an evidence-based program or practice (EBP), to determine whether the program or practice is implemented according to design specifications. When an organization says it is using an EBP, it is necessary to implement the EBP with "fidelity" in order to expect the same results that have been shown to occur through previous research. Many EBP creators have developed materials to help organizations know what needs to be done to adequately comply with the requirements of that EBP. If the program designers have not done this, you will need to create them for your own use.

Questions related to fidelity assessment have a particular importance when using an EBP (refer back to chapter 6 regarding finding and selecting EBPs). Every EBP must be assessed to indicate how well plans were followed for that particular service, particularly in terms of how well the core components were implemented. This is similar to the idea of the process evaluation based on the logic model, but particularized to the program or practice. The implementation of the EBP can be expected to lead to the positive outcomes noted in research only if it is implemented properly. Thus, the evaluation plan must show how the program will be monitored and assessed for fidelity. Of particular importance is the idea of noting whatever deviations occurred and determining why the deviations occurred and with what effect. Not all differences between plan and reality are negative. Some variations are in response to the environment and client population needs, and these types of adaptations can lead to progress in applying the EBP in appropriate new ways.

OUTCOME EVALUATION

Outcome evaluations seek to determine whether the outcomes desired by the program for client change are achieved. (This type of evaluation is also called "summative evaluation.") Outcome evaluations are able to assess the worth of the program in terms of whether clients change and their problems are reduced or solved. We should be able to say whether the problem stated in the logic model has been reduced or eliminated after conducting an outcome evaluation. Outcome evaluations determine the extent to which the outcomes listed on the logic model are accomplished (see chapter 7). Outcome evaluation focuses on the columns on the right side of the logic model, where the designated short-, medium-, and long-term outcomes are listed. (Remember that sometimes these outcomes are still referred to as "goals and objectives.")

With some programs and evaluations, it is impossible or undesirable to measure *all* of the outcomes. Thus, choices need to be made. The primary reason that outcomes are not measured is that they are not thought to be achievable immediately or within the scope of the program's funding, which often is only for a few years. For example, working with students to improve their learning abilities while they are in kindergarten may improve high school graduation rates and life satisfaction, but those outcomes are so far in the future that grant funding will almost certainly be ended before they can be measured. Unless long-term funding is in place to continue an evaluation, it is unlikely that these very long-term outcomes will be assessed. Another reason not all outcomes are measured in an evaluation is that there are a large number of outcomes and, given funding levels, there are not enough resources to track all of them.

It is important, as you plan the outcome evaluation, to also be planning how the information will be used to judge the value of the program. This judgment is made in the context of a comparison. One type of comparison can be made between your clients at the start of the program and your clients after the program is completed for them. (This is called a pre-test/post-test research design.) Another type of comparison is between your clients after the program, and another group of similar people who did not receive the program. (This is called a comparison group design.) If you combine these approaches into a pre-test/post-test comparison group design, you have a good basis to conclude whether the program was truly effective or not. If you are able to assign clients randomly either to the treatment group or the comparison group, your evaluation design becomes quite strong. While a full discussion of research design in all its nuances is beyond the scope of this book, it is useful to know that most funders accept a pre-test/post-test approach to generating evaluation results. If a higher standard is required, the request for proposals will certainly make that clear.

MEASUREMENT IN EVALUATION

Measurement is an important part of any evaluation effort. The evaluation plan chooses and then describes which aspects of the program (process, fidelity, and outcome) to assess from among the various outcomes in the logic model. The logic model provides a quick overview of which outcomes are available to measure (*what* to measure) but decisions must also be made about *how* to measure the outcomes. It is up to the grant writer, with the possible aid of an evaluation specialist or consultant, to determine how each of these outcomes can be measured. The value of the evaluation will be seriously compromised if measures are not appropriate or have low validity and reliability. It is suggested that anyone designing an evaluation look at a book on research methods such as Rubin and Babbie (2012), and also have access to books about measures, such as Fischer and Corcoran (2007). (The cost of a new book on research methods may be pretty high, but used editions contain much the same information and can be found for much lower prices.)

Measurement can be quantitative (numerical) or qualitative (verbal) and frequently combines both approaches. One of the qualities of good outcome evaluations is that they use more than just one way to measure important concepts. The best way to accurately determine the achievement level of outcomes is through using multiple sources of information (teachers, students, case managers, parents) and multiple reporting formats (quantitative, such as a standardized instrument or some official school records, and qualitative, such as journal entries and interviews with important stakeholders, such as parents and teachers).

MEASUREMENT TRADE-OFFS

As you plan which measurement instruments to use, you must make allowances for at least two different trade-offs. The first is the ability to accurately measure outcomes (perhaps using multiple measures of the outcome) versus the use of a "pretty good" or "good enough" measure of the outcome. More elaborate measurement approaches will yield more valid or accurate information but will usually take longer and cost more than measurement approaches that are less precise.

The second trade-off is between "pure" laboratory-like collection procedures that are as standardized as possible, versus practical procedures that obtain any useful data. For example, it is not ideal to have a case worker collect information directly from clients about their satisfaction with the program—it is better to have a neutral observer do this. But clients don't necessarily want to share possibly sensitive information with a stranger such as an evaluator. Clients also often do not want to have a special meeting for data collection once their involvement in the intervention is over. They may therefore resist this sort of post-program involvement. If clients do not provide their information and insights, very little data may be collected. As you design your data collection process, be mindful of client preferences as well as research rigor.

STANDARDIZED MEASURES AND OUTPUTS

When you choose how to measure aspects of the program (particularly client outcomes) it is best if you can use standardized instruments because such tools have properties that are already known, such as the average score for one or more groups, and acceptable levels of validity and reliability. This makes it easier to make comparisons between your clients and themselves at a different time, or between your clients and another group.

Sometimes, however, it can be difficult to find a standardized instrument that is fully appropriate and relevant to your outcome. In such cases you will need to design your own measurement instruments. Creating your own instrument has the advantage of simplicity and of being directly connected to your evaluation. The downside of developing something just for your situation is that measures that will be accepted by the funder are not necessarily easy to come up with. A considerable amount of knowledge (and time) can be required. Much depends on the nature of the concept to be measured.

Another element that is frequently overlooked is using standardized definitions for outputs rather than developing your own, ad hoc, definitions. For example, suppose you have as a program activity that you will provide counseling to youth. An output of counseling is the provision of "one session" for "one youth." Typically, in human services, the standard unit of time for an educational session is one hour. If your program, in seeking to increase the number of outputs, changed this standard definition without notice to only 30 minutes, the program would look a lot better in terms of numbers, but youth would be shortchanged and a funder might feel as if the program were acting unethically. In short, use standard definitions of outputs (and other aspects of the program) when available and always be clear in your definition of terms.

USING LOGIC MODELS IN EVALUATION

As you have already learned, we can use a logic model to represent what we believe will happen when the proper inputs are applied to the correct client population (Frechtling, 2007; see chapter 7). In the end, if all goes well, clients will no longer have the problem the program addresses or will have it to a lesser degree. Using your program's logic model helps to quickly plan the process and outcome evaluation components of your proposal.

USING THE LOGIC MODEL IN PROCESS EVALUATION

The value of the logic model for process evaluation is that most of the conceptual information needed to design the evaluation of a program is present. The required inputs are listed, and the evaluator can check to determine which resources actually came into the program. Activities are similarly delineated, and an evaluator can usually find a way to count the number of activities that the program completed. Similarly, the logic model describes what outputs are expected, and the evaluator merely has to determine how to count the number of completed outputs that result from the program activities.

Looking at the example logic model (Figure 8.1, repeated from chapter 7) shows us that we want to have in our evaluation plan at least one way to measure whether funding, staff, and space (the inputs) are adequate; how much case management occurred and individual counseling was conducted (the activities); the extent to which referrals were made (and

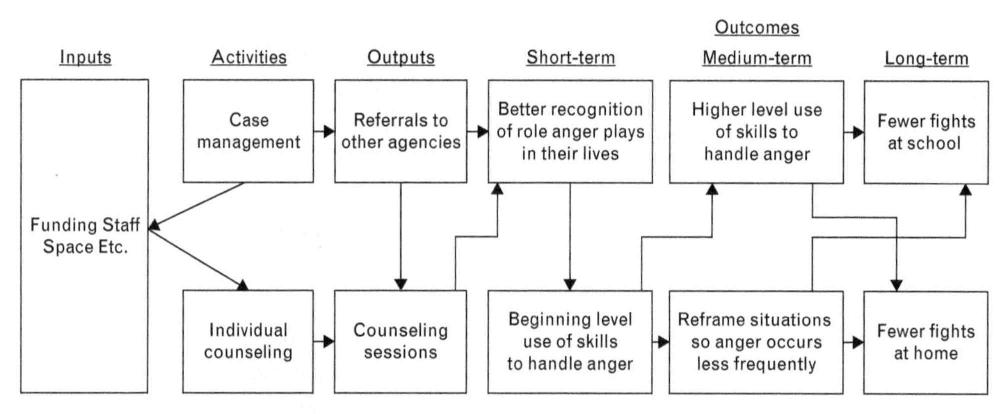

FIGURE 8.1: Example Logic Model (from chapter 7)

followed up on); and the number of individual counseling sessions that happened (the outputs). This information should be in program documents to compare what was planned for with what was actually provided. Having a logic model from the beginning allows the evaluator to ensure that proper data are being collected from the program's start, rather than scrambling later to answer some of these basic questions.

Among the easier elements of designing the process evaluation plan is to describe which resources have been provided. The evaluative function should describe what these numbers are and should also evaluate their adequacy. The first of these tasks is straightforward; the second is more difficult. Program records will show the amount of resources expended on behalf of the program, the number of staff members initially hired, the space allocated to various components of the program, and so on. But having only these numbers may not be enough to judge adequacy.

The need to discuss adequacy relies on a clear statement in the grant application of the *rationale* for particular levels of resources. For example, if only MSW-level social workers with a clinical license are able to provide services properly, this needs to be explained under program plans. Then, when you are evaluating the adequacy of staff members for the job, you can note that all people hired had this credential. If not everyone did, then this would be a case of needing to point out a gap between declared need and actual implementation. The same is true for space—there may be regulations about how much space is required for the type of service you are proposing, so a lack of adequate space is a very serious issue. For now, though, in the grant application where you explain your evaluation plan, it is enough to state that you will compare resources obtained with the level and type of resources set forth in the program plan.

As noted earlier, this is not a perfect logic model. The question in the process evaluation at this stage might be to determine how to actually measure "case management." The output is supposed to be "referrals to other agencies," but there is much else that could be considered beneficial from a case management approach. This element may need careful delineation and discussion with stakeholders as you develop the grant application to ascertain exactly what is important about case management that should be measured.

Let's think about "case management" for a bit. According to the Case Management Society of America (n.d.), "Case management is a collaborative process of assessment, planning, facilitation, care coordination, evaluation, and advocacy for options and services to meet an individual's and family's comprehensive health needs through communication and available resources to promote quality, cost-effective outcomes." If this is the definition your organization chooses to use, a deep assessment of whether case management is occurring would have to develop a measure that looks at all of these aspects, looking for evidence that the case manager is engaged in assessment, planning, facilitation and all the other aspects of case management, and doing so in a "collaborative" way.

As the grantwriter, you must work with the program staff, who may already be employed by the agency, to ensure that your plans for measuring a particular term make sense to the people who will be evaluated in terms of whether they are doing the job they were hired to do. If you do not work with other stakeholders on your plans, it is possible that you'll create a "wonderful" measurement system that is so complicated that it is soon abandoned. (See Table 8.1 for an example of a tool to measure "case management.")

TABLE 8.1: Hypothetical Measurement Tool for Case Management

Client ID #: ________	To what extent did the Case Manager complete these tasks collaboratively, through communication and the use of available resources, for THIS Client over the past 30 days?				
	Very Much	Much	Some	Little	Very Little
Assessment					
Planning					
Facilitation					
Care Coordination					
Evaluation					
Advocacy					

The question of whether this represents too much intrusion into the daily life of the worker is an open one. Measures that take too much time to complete are likely to be resented and possibly not filled in meaningfully. Again, a trade-off must be considered in the planning of the evaluation. Staff members will likely explain they are too busy to fill in what they may see as "meaningless paperwork," while funders want evidence that activities are being conducted properly and outcomes are being achieved at reasonable levels. If the funder feels the money is not well used, there won't be any staff members at all!

It is important to also note that all is not lost if it turns out that reality and the plan did not match. Much can be learned from deviations from the plan that may be important lessons. For example, it might be that the case management definition adopted for this program wasn't implemented well because of constraints of the school district and its policies. These challenges should be noted and adjusted to, and should be explained to the funder. This type of finding can be useful to other grantees as well. In general, being able to indicate *who* provided *what* services to *which* clients is the most important indication of a job well done with the process evaluation.

USING A LOGIC MODEL IN OUTCOME EVALUATION

Just as the logic model is very helpful in designing the process evaluation, it is also a great aid in setting up the outcome evaluation (Frechtling, 2007). But decisions need to be made about how to define and then measure the outcomes that are listed here as well. Let's take the example of the outcome of "recognizing the role anger plays." You could measure this in at least two ways. First, you could request a statement from the client indicating that the client has "recognized the role anger plays" in his or her life, without going into any detail. A second approach would be to have the case worker or counselor write a statement about the role anger plays in the client's life. Neither of these measurements will have a lot of practical utility. The client may very well agree to write a short "Yes I recognize the role anger plays in my life" statement without it being true or showing insight. Asking another person (such as the case manager) to write a statement about the client may not prove insightful, either. Going through the logic model in this way actually shows that this link in program logic is difficult to measure and may not be necessary.

While it may seem startling to have an example in a text that shows a less-than-perfect approach, it is included here to show that using a logic model is very useful in showing weak spots in the program logic. This outcome, "recognizing the role anger plays," and the links from the outputs and to other outcomes are not perfect but are not necessarily fatally flawed, either. The issue for evaluation is how to measure it, and whether it really needs to be measured at all. Going through the proposed program with an evaluator's perspective is helpful to improve the program before the proposal is even submitted (Box 8.1).

In the example logic model, the short-term outcome "better recognition of the role anger plays in their lives" must be measured because it is the foundation for all other outcomes. The measurement plan could be developed using a set of questions asked at intake into the program and after some time has passed after receiving services. One standardized anger management instrument is called the "Anger Management Scale" (Stith & Hamby, 2002). A standardized instrument such as this, if it is appropriate for the clients and program, is a good choice because you can find norms, or expected response levels, for the items on the instrument. It also indicates you have done your homework in terms of understanding how your program's evaluation can be connected to previous research on the topic.

Several other possibilities exist for measuring the outcome of "recognizing the role anger plays." A way to measure this concept would be having the students keep a daily journal of the times when anger (or its precursor) was felt and how it was handled. The journal could be examined by the case manager and evaluator to look for evidence of changes in level

BOX 8.1: UNANTICIPATED OUTCOMES IN EVALUATIONS

Outcome evaluations also sometimes include a search for unanticipated outcomes. An unanticipated outcome is a change in clients or the environment that occurs because of the program, intervention, or policy no one hypothesized would result. Thus it is not included in the logic model. Unanticipated outcomes (whether positive or negative) are frequently not discovered because no one is looking for them. Qualitative approaches to measuring program effects may be more likely to uncover unanticipated outcomes because they are more open-ended in nature. This allows for information about new topics to be found. It is thus often a good idea to include some interviews with clients, staff members, and others affected by the program in order to capture data that would otherwise not be included in the results.

of insight. Teachers and parents could also be asked for their insights. These are primarily qualitative measures. More objective (quantitative) indicators such as number of times a student was written up for disciplinary actions due to expressing anger can also be combined with the subjective judgments of participants and other stakeholders.

Journal entries could also note the use of techniques to handle anger, which needs to be measured as another short-term outcome as well as being a medium-term outcome. A journal could actually be a qualitative way to measure the other outcomes at the medium and long-term levels and so serve multiple purposes.

Also of importance is the next link in the logic model, which leads to "learn skills to handle anger." The evaluation must ensure that clients understand skills to help them handle anger and so document these skills. It is not enough to indicate that skills were taught, as in a group or individual session. Teaching a class is an activity, and so would be documented in the process evaluation portion of the overall evaluation, but being in a class does not guarantee a change in the client. In this evaluation, we would like to have a measure of skills that shows improvement in the ability of the students to use the anger management skills being taught. This attribute of the measure is important because we expect the clients to get better in their use over time and include more skillful use of the techniques as a medium-term outcome in the logic model.

The other medium-term outcome expected is that clients will be able to reframe situations so that they actually get angry less frequently. The program logic shows this outcome occurring as a result of both beginning and higher level use of skills. Because this element is broken out from the use of skills to "handle anger," it will need a separate measure. As an evaluator, you can hope that an established, normed instrument is available, or that this is a skill that is measured by a separate item on a longer scale. If not, you will need to find a way to pull this information from staff members' reports or client self-assessments.

The final links in the logic model connect the medium-term outcomes to the long-term outcomes, which are fewer fights at school and fewer fights at home. Because youth having too many fights was identified as the problem this program is addressing, we want to know to what extent fights decreased. The measure here could be client self-reports, school records, and/or reports from people living in the home.

OUTCOME EVALUATION DATA COLLECTION PLAN CHART

One of the most useful ways to understand the evaluation plan is to create a chart with the outcome to be measured in the first column, how it will be measured in a second column, and columns for "who will collect" and "when collection will occur," as shown in Table 8.2. It is important to note that outcomes can be measured in more than one way. Multiple measures are a good idea in an outcome evaluation.

The data collection process is presented extremely clearly with an "Outcome Evaluation Data Collection Plan" chart. The outcome is laid out, the actual measure is described, the person responsible is named, and the frequency is clearly set. It does not take much time to create the plan and lay it out in this manner.

Table 8.2: Example Outcome Evaluation Plan Data Collection Chart

Outcome	How Measured?	Who Will Collect	When Collection Occurs
Better recognition of role anger plays in student's life	Student writes in a journal every day about feelings and how handled	Case manager collects journal entries	Every Monday morning; to be returned to student by Monday afternoon
Better recognition of role anger plays in student's life	Student completes the "Anger Management" scale	Case manager	On first day of program, and every four weeks until end of program
Beginning level use of skills to handle anger	Student completes the "Anger Management" scale	Case manager	On first day of program, and every four weeks until end of program
Beginning level use of skills to handle anger	Student's teachers fill in form weekly noting if student appears to become angry and how it was handled	Case manager collects from teachers	Every Friday for each student in program
Beginning level use of skills to handle anger	Student's parent or parents fill in form noting if student appears to become angry and how it was handled	Case manager collects from student, who brings in statement from parent	Every Monday during program
Higher level use of skills to handle anger	Student writes in a journal every day about feelings and how handled	Case manager collects journal entries	Every Monday morning; to be returned to student by Monday afternoon
Higher level use of skills to handle anger	Student completes the "Anger Management" scale	Case manager	On first day of program, and every four weeks until end of program
Higher level use of skills to handle anger	Student's teachers fill in form weekly noting if student appears to become angry and how it was handled	Case manager collects from teachers	Every Friday for each student in program
Higher level use of skills to handle anger	Student's parent or parents fill in form noting if student appears to become angry and how it was handled	Case manager collects from student, who brings in statement from parent	Every Monday during program
Reframe situations so anger occurs less often	Student journal entries are asked to address if reframing happened during the day and with what results	Case manager collects journal entries	Every Monday morning; to be returned to student by Monday afternoon
Fewer fights at school	School disciplinary reports	Case manager	Every week
Fewer fights at home	Reports from parent	Case manager collects from student who brings in statement from parent	Every week

WHAT DOES THE EVALUATION SECTION OF A GRANT PROPOSAL NEED TO COVER?

The answer to the question of what needs to be covered in the evaluation section of a grant proposal for *your* grant is whatever is in the announcement *you* are responding to (of course). Still, because requirements vary considerably, two examples are provided to show the diversity of what you may encounter as you develop this section.

EXAMPLE 1

The first example is from the US Department of Health and Human Services, Health Resources and Services Administration, HIV/AIDS Bureau, Special Projects of National Significance Program, HRSA-12-100. (See Table 8.2.) For this grant proposal, the evaluation section was worth 20 points out of 100.

The first paragraph of the evaluation capacity section of the request for proposals (RFP; see Box 8.2) tells the reader exactly what to do. A grantwriter merely needs to write two statements agreeing to participate in the multisite evaluation and to submit proof of Institutional Review Board (IRB) approval to the funder. As an example of what you could write to comply with the first criterion, you could submit this one sentence: "The applicant will work closely with the Evaluation and Technical Assistance Center and fully participate in the multistate evaluation activities that are developed, such as data collection and reporting of outcome, process, and cost data."

Unfortunately, sometimes grantwriters do not follow directions. Reviewers are then required to deduct points from that applicant's possible points. It is thus very important to examine closely the description of the required information as well as the criteria for receiving points. Because requirements can be scattered throughout the RFP, you need to read it several times and note all information needed before you begin to write your proposal.

The second paragraph is also clear on what is requested. Using key personnel's resumes, the grantwriter can develop the required description. This section will be more difficult to write if the people who will be in these positions are not yet hired, but you can nonetheless describe the skills and experience you will want in the people who take on these roles.

The third paragraph discusses a potential local evaluation. While the applicant *must* agree to be a part of the multisite evaluation, the RFP does not require a local evaluation. Thus, the grant writer does not need to do anything here, if that is the grant-seeking organization's choice. There may be, however, an unstated expectation that a local evaluation will be conducted. This brings up the point that not only is it important to read the RFP carefully but also you should seek to collect additional information, such as by speaking with the contact person for the grant proposal process, attending bidders' conferences (or at least listening in via phone or webinar connection), and so on. It may be assumed by the funder that local evaluations, while not required, are an indication of an organization that is more

BOX 8.2: REQUIREMENTS FOR EVALUATION SECTION FOR THE SPECIAL PROJECTS OF NATIONAL SIGNIFICANCE PROGRAM, HRSA-12-100

State explicitly your willingness to participate in a five-year comprehensive multi-site evaluation and to fully cooperate with the Evaluation and Technical Assistance Center (ETAC) throughout the initiative. This cooperation includes but is not limited to data collection and reporting of outcome, process and cost data for the multi-site evaluation and additional focused evaluation studies; and publication and dissemination efforts of the initiative's findings and lessons learned at the national, State and local levels. State your agreement to submit proof of IRB approvals and renewals for all client-level data collection instruments, informed consents and evaluation materials to SPNS and to the ETAC on an annual basis.

Describe the prior experience of proposed key project personnel (including any consultants and subcontractors) in participating in a multi-site evaluation of national scope. Describe the experience of proposed key project personnel (including any consultants and subcontractors) in writing and publishing study findings in peer-reviewed journals and in disseminating findings to local communities, national conferences and to policy makers.

If a local evaluation is included by the applicant, describe the organization's capacity to conduct it. Describe the local evaluation plan with proposed outcome measures that will demonstrate whether the intervention's goals and objectives are met. Describe how the proposed key project personnel (including any consultants and subcontractors) have the necessary knowledge, experience, training and skills in designing and implementing public health program evaluations, specifically evaluations of innovative HIV access and retention projects. Include any specific experience in the evaluation of programs serving HIV positive homeless or unstably housed individuals with co-occurring mental illness or substance abuse issues. If applicable, detail any published materials, presentations and previous work of a similar nature.

Identify the Institutional Review Board (IRB) which will review the multi-site evaluation plan and, if applicable, the local evaluation plan. Describe any training in human subjects research protection by proposed key project staff. Describe your written plan to safeguard patients' privacy and confidentiality, and your documented procedures for electronically and physically protecting the privacy of patient information and data, in accordance with HIPAA regulations and human subjects research protections (p. 29).

serious, has better qualifications, or other attributes of a "good" grantee. You won't know this unless you make extra efforts to research the issue.

Assuming you decide to plan a local evaluation, you would naturally include the requested information. "Capacity to conduct a local evaluation" is primarily about the staff members or consultants with adequate skills and experience in leading evaluations as well as dedicating sufficient resources from the grant budget. Reviewers will assess the strength of the local evaluation plan as well as the staff capacity, so bringing in a well-qualified evaluator who can create a methodologically rigorous evaluation plan is a frequently used approach. In most cases, it is very helpful to have the person who would conduct the evaluation write this section of the grant proposal.

The final paragraph asks for information relating to protecting human subjects. An IRB is a standard safeguard for any program conducting research that involves human subjects. All universities have one, and if your evaluator is a contracted person who also works at an institution of higher education, your evaluator may be able to submit and gain approval through his or her university. Other IRBs exist as well that will examine your human subjects protection plans for a fee, and approve if all is in order. The RFP review criteria don't ask you to have already received approval. They want you to state in the application which IRB you will submit to and that you agree to provide documentation that you will provide the funder with information about your organization's IRB status. Both of these elements are very easy to include; not doing so will mean points will be deducted from your score. These lost points may be the difference between receiving the grant and not.

EXAMPLE 2

The second example of the information in the request for applications (RFA) regarding the evaluation section comes from the "Administration on Children and Youth's Grants to Trafficking Within the Child Welfare Population Program (HHS-2014-ACF-ACYF-CA-0831) Request for Applications" (ACYF, 2014). Early in the RFA it is made clear that the Child Welfare Trafficking Grants (CWTG) take evaluation seriously.

> CWTG projects are required to develop and implement projects and activities that are outcome-focused and include measurable objectives or steps that can measure progress in meeting desired outcomes. They must implement an evaluation plan that measures progress and identifies evidence of the project's impact and success. (Administration on Children and Youth, 2014, p. 4)

This RFA has much more emphasis on the details on the evaluation plan than does the RFA in Example 1. That is because in Example 1 most of the information that the funder wants to gather will be collected through a multisite evaluation effort that is controlled outside the confines of the individual grant recipient. The grantee just has to agree to do what is requested by the evaluation team at the national level. In this second example, the emphasis is on the local organization coming up with its own evaluation approach. The grantee has to be fully in charge of its own efforts, so the grantwriter must provide more information to do well in the competition for funds than in the previous example. Still, it is remarkable that the RFA tells the careful reader exactly what needs to be written in this section.

The authors of the RFA used here in the second example divide the information about evaluation into two parts. The first is called "Program Performance Evaluation" while the second is called "Funded Activities Evaluation." At first glance, there is a lot of overlap between these two example sections of the evaluation (or "program performance evaluation") requirements. Some of the language is very similar. You must keep in mind, however, the different contexts for the two subsections: the first is "program performance evaluation" while the second is "funded activities evaluation." The first is more about process, while the second is more about outcome. (See Box 8.3 and Box 8.4 for the text of these two components of the evaluation expectations described in the RFA.) Each sentence contains important information. While most of this is straightforward, sometimes it takes some time to decode the RFA's intent.

Note in the first paragraph that the evaluation "will contribute to continuous quality improvement." While there are many definitions of "continuous quality improvement" (CQI), a leading foundation defines the term in this way: "Continuous quality improvement (CQI) is the process-based, data-driven approach to improving the quality of a product or service" (Robert Wood Johnson Foundation, n.d.). An underlying belief is that it is always possible to improve the way things operate and to improve quality. Thus, the program

BOX 8.3: PROGRAM PERFORMANCE EVALUATION PLAN

Applicants must describe the plan for the program performance evaluation that will contribute to continuous quality improvement. The program performance evaluation should monitor ongoing processes and the progress towards the goals and objectives of the project. Include descriptions of the inputs (e.g., organizational profile, collaborative partners, key staff, budget, and other resources), key processes, and expected outcomes of the funded activities. The plan may be supported by a logic model and must explain how the inputs, processes, and outcomes will be measured, and how the resulting information will be used to inform improvement of funded activities.

Applicants must describe the systems and processes that will support the organization's performance management requirements through effective tracking of performance outcomes, including a description of how the organization will collect and manage data (e.g., assigned skilled staff, data management software) in a way that allows for accurate and timely reporting of performance outcomes. Applicants must describe any potential obstacles for implementing the program performance evaluation and how those obstacles will be addressed (p. 20).

performance evaluation described in this RFP must be designed to operate to understand and improve the processes of the grant-funded program.

The remainder of the first paragraph presents specific elements that must be included in the narrative for the section. Note the discussion of how the logic model elements (inputs, processes/activities, and outcomes) must be measured and also how those measures will be used to improve the program.

The second paragraph tells the reader to showcase how performance outcomes will be tracked, measured, and used for decision-making purposes in a timely way. While it may take more than one draft to get all this information included, the information in this chapter should help you make sense of all these terms and requirements.

Moving on to the second part of the evaluation plan, we come to the "Funded Activities Evaluation" section of the RFA (see Box 8.4). As in the previous section, the information contains a great amount of detail that needs to be taken into account. The more terms and ideas described in the RFA you can include in your proposal, the more likely you will be to receive a high score on your efforts.

Evaluating funded activities is the evaluation component that assesses how close the project has come to achieving its goals and objectives/expected effects/impacts/outcomes. This part of the RFP is somewhat difficult to untangle but every grantwriter should work as many phrases into the narrative as possible. You need to create and explain underlying connections related to your logic model and use the other tools provided earlier in the chapter. The key aspect of any evaluation plan is to make sure that you are telling the potential funder what data are needed, how you are going to collect that data, and

BOX 8.4: FUNDED ACTIVITIES EVALUATION

Applicants must describe the plan for rigorous evaluation of funded activities. The evaluation may be supported by a logic model. The evaluation must assess processes and progress towards the goals and objectives of the project, and whether the project is having the expected effects and impacts. The evaluation plan must specify expected outcomes and any research questions. The plan must discuss how the results of this evaluation will provide greater understanding and improvement of the funded activities. The plan must include a valid and reliable measurement plan and sound methodological design. Details regarding the proposed data collection activities, the participants, and data management, and analyses plans must be described. Applicants must describe any potential obstacles foreseen in implementation of the planned evaluation and how those obstacles will be addressed.

Applicants must describe how their evaluation plan addresses each of the goals and objectives, and builds upon existing literature as referenced in *Section I*. Additionally, applicants must describe and provide a rationale for the data they plan to collect (p. 20).

how you are going to use the information collected to determine whether the program is "working" or not.

As an interesting aside, this section seems to indicate that the use of a logic model to design the evaluation is a "good idea" but not necessary. Looking at the criteria given to the review panel for scoring, however, one may see that the proposals are being judged (and having points deducted) if there is no logic model in the submission. This situation is another indication just how important it is to read the ***entire*** RFP before beginning to write the proposal.

SUMMARY

This chapter has laid out reasons for paying significant attention to the evaluation or performance assessment section of the grant proposal. It is typically worth 15 to 25 percent of the entire set of points, so achieving a high score in this section is often the difference between being successfully funded or not. The evaluation section is exceedingly important to the success of the grant proposal. This fact, when combined with the truth that the knowledge needed to do an excellent job in the evaluation role is rare, means that it can make sense to bring in an evaluator from outside the agency to write this section. Still, if it is not possible to bring in assistance, this chapter will help any grantwriter know what to write in this section:

- Pay close attention to what the RFP says is needed in the evaluation section (both in the description of the section and the points criteria section),
- Use your logic model to guide both the process and outcome aspects of the evaluation, and
- Develop an "Outcome Evaluation Data Collection Plan" chart showing how each important outcome of the program will be assessed, as shown in Table 8.2.

PRACTICE WHAT YOU LEARNED

1. Looking at your logic model, using the requirements of your RFP, and taking into account all of the information in this chapter, create a parallel to Table 8.2 for your program. Explain which concepts you are involving in your plan, determine how you will measure each one, state who will collect the information and explain how often it will be collected. Remember that you may be required by the terms of the RFP to collect certain information, while other information is up to you to include or not.
2. Begin to write up the evaluation section for an RFA you are interested in using or you have found from previous funding cycles. Share this with a colleague who has experience in grantwriting, or whose opinion you trust. Get critical feedback from that person and make suggested changes.

REFERENCES

Administration on Child, Youth and Families (2014). Grants to Address Trafficking Within the Child Welfare Population (HHS-2014-ACF-ACYF-CA-0831). Retrieved from www.acf.hhs.gov/grants/open/foa/file/HHS-2014-ACF-ACYF-CA-0831_0.pdf

Case Management Society of America (n.d.). What Is a Case Manager? Retrieved from http://www.cmsa.org/Home/CMSA/WhatisaCaseManager/tabid/224/Default.aspx

Fischer, J. & Corcoran, K. (2007). *Measures for Clinical Practice and Research: A Sourcebook* (4th ed.). New York: Oxford University Press.

Frechtling, J. (2007). *Logic Modeling Methods in Program Evaluation.* San Francisco: Jossey-Bass.

Robert Wood Johnson Foundation (n.d.). The Science of Continuous Quality Improvement. Retrieved from www.rwjf.org/en/research-publications/research-features/evaluating-CQI.html

Rossi, P., Lipsey, M., & Freeman, H. (2003). *Evaluation: A Systematic Approach* (7th ed.). Thousand Oaks, CA: Sage.

Rubin, A. & Babbie, E. (2012). *Essential Research Methods for Social Work* (3rd ed.) Belmont, CA: Brooks-Cole.

Stith, S. & Hamby, S. (2002). The Anger Management Scale: Development and preliminary psychometric properties. *Violence and Victims, 17,* 383–402.

Substance Abuse and Mental Health Services Administration (SAMHSA) (n.d.). Using Process Evaluation to Monitor Program Implementation. Retrieved from http://captus.samhsa.gov/access-resources/using-process-evaluation-monitor-program-implementation

CHAPTER 9

PROGRAM IMPLEMENTATION PLANNING

Once you have discussed the need your community has, chosen or developed a program that has evidence to support its efficacy, drawn a logic model of that program, and created an evaluation plan, you have come quite far in writing your proposal. You might think that the main work is now done. But, in fact, all the work you've put in so far as a grant writer is not sufficient. If your proposal is chosen (and why shouldn't it be?) the funder is going to want to know more before handing you a check for operating expenses. The funder is going to want to know that you can actually do what you promise to do. In this part of the proposal, you must show that you have a plan to put your proposal into action quickly. The funder may take a long time to get a decision to you, but once you've been chosen, you must be ready to move. In a way, you need to have this project "shovel ready," which means that you need to know what type of person to hire, how much to pay, how much activities will cost, and a host of other details that you might think could wait until you know you've been chosen. But in this Age of Scarcity, you don't want to be left behind in the race to attract resources. One of the key differences between "almost ready" and "really ready" is that those who are "really ready" have a plan they can share with their funder at the time the request for resources is submitted while the "almost ready" are hoping for more time to "get ready." Another difference is that the "really ready" get funded and the "almost ready" don't.

This chapter takes you through the process of understanding what you need to know in order to effectively share the plans you have for implementing your proposed program in the grant request. By the end of the chapter, you should be able to write a comprehensive description of how you will be able to move forward quickly to serve clients as soon as the money starts flowing. This is a challenging aspect of any grant proposal, because you must have a very specific vision of the program you are asking resources for—who is going to do what, when, how often, and how you (as an agency) are going to move from your current state to having a fully operating new program up and running.

WHAT IS PROGRAM IMPLEMENTATION PLANNING?

It has often been said that "those who fail to plan, plan to fail." An implementation plan of your program lays out the steps and milestones by which you will start and operate the program you are proposing. Funders often (but not always) allow some time for a "start-up" period. By the end of this time period, your staff must be hired and trained, necessary supplies must be on hand, clients must be recruited, protocols for cooperative agreements must be on file, and you must actually be assisting the population you wrote the grant for.

The benefits of an implementation plan are many. Roper, Hall, and White (2011) believe that an implementation plan is a useful management tool that acts as a map to illustrate critical steps when beginning a new program or project. A clear plan assists in identifying potential issues early on before receiving the grant, allowing the organization to be proactive rather than reactive. An implementation plan forces you to do a "walk-through" from receiving the go-ahead from the funder to providing services to recipients. As you plan, you are likely to anticipate challenges to successful implementation and operation that you wouldn't have considered beforehand. For example, recruiting and hiring qualified staff members is not always as easy as putting an advertisement in the paper. If you need people with special qualifications, such as fluency in a language other than English, or who live in a certain place (particular neighborhoods in urban centers or rural areas), you might have difficulty in finding appropriate candidates for the amount of money you want to pay. Accordingly, your implementation plan will need to reflect the extra steps you will take in the hiring process.

Recruitment of clients may also need to be planned in more detail than you at first anticipate. If your agency is requesting funding to assist a very particular type of client, you may need to set up additional means to attract such people to your program. An agency I once worked with wanted to expand their program to provide services to homeless, mentally ill, substance-abusing people, particularly those who were Hispanic. A great deal of thought went into the processes by which they would work with other agencies to receive referrals of potential clients and how the agency would react once a referral was made. Their client recruitment processes were reimagined from the vantage point of attracting that type of client. This plan helped the agency receive a grant worth over $1,000,000 in total from the federal government.

Another benefit of having a clear implementation plan is that it helps staff members share a common idea of how the program will begin and be run. The anticipated activities, outputs, and outcomes of the program are embedded in the plan, and everyone clearly understands how the plan will proceed. Issues can be examined early on, rather than being discovered only once the program commences.

An implementation plan is a guide for *developing* the program, as well as for *running* it. It allows you time to pilot procedures, strategies, and techniques to ensure that they are appropriate in your situation and with your clients. These should all be finalized by the end of the start-up period. One way to think of this process is as if you are opening a store: before the doors open for business, you need to make sure the products are on the shelves, you must have a sales team with training about the products waiting to assist customers, and you should have recruited customers so they are coming in the door to buy as soon as the doors are unlocked. This start-up process is not the same as running the store once it is open, but many of the decisions you make during start-up will have a large impact on the day-to-day operations once your program is operating full-force. The start-up plan just gets you to opening day, but its impact is long-lasting.

A final benefit of creating a comprehensive implementation plan before you receive funding is that you can then put the plan into effect instead of having to think through all the details later, when time is running and clients are waiting. Sometimes, before you actually receive the funds you were awarded, the funder will send a team of people to visit your organization on a site visit. This team is there to ensure that you're able to do everything you've written into your proposal. There is enough pressure during this site visit that you don't want to still be in the development stages of your plan. This is not to say that you won't be able to make modifications as the plan is used. You may find that some program elements were not thought of, despite your best efforts, and so must be developed on the fly. You may also face unanticipated challenges that need to be handled. Some of your initial ideas may not actually work out. But you are much farther along in the implementation process with a well-developed plan than if you did not have a plan at all.

Of course, perhaps the final reason to develop a sound implementation plan is that it is required by the funder to include in the original proposal. While the percentage of points it is worth varies from one funding opportunity to another, it is not uncommon for the implementation plan to be worth 20 to 30 percent of all points awarded.

WHAT SHOULD BE IN THE IMPLEMENTATION PLAN?

While each grant application will be somewhat different (you must always look at what is being requested by *your* potential funder), there are common elements of program implementation plans. Use this section as a general roadmap to the information needed and refer to your funder's requirements about the topics that must be in *your* implementation plan.

The following information (derived from Roper et al., 2011) should be provided for each site where a program element is to be implemented. This list is broken into subcategories: General information; site-specific information (when your program is implemented in more than one location); staffing; client recruitment; fidelity assessment; data collection and evaluation; and referral system (for more detail, see Roper et al., 2011). If you include all this information, you will have an excellent implementation plan. Again, it is vital to include what is in the Request for Applications (RFA) you are applying for.

EXAMPLE OF A FEDERAL GRANT IMPLEMENTATION PLAN SECTION

The following section taken from a federal grant will seem shockingly short. Still, the amount of information that is being requested is considerable, compared to the question being stated. The section reads: "Describe how the proposed activities will be implemented and how adherence to the National Standards for Culturally and Linguistic Appropriate Services (CLAS) in Health and Health Care will be monitored" (SAMHSA, 2013, p. 25). An additional bullet point in this section says, "Provide a chart, graph, and/or table depicting a realistic timeline for the entire 5-year project period, showing key activities, milestones, and responsible staff" (SAMHSA, 2013, p. 25).

Other federal requests for proposals in the past have been much more detailed, with a page or two of explicit directions, even from the same agency (SAMHSA). A small portion of those directions is included in Box 9.1. While there is a lot of detail in the following

BOX 9.1: SECTION C OF SAMHSA'STARGETED CAPACITY EXPANSION PROGRAM FOR SUBSTANCE ABUSE,TREATMENT AND HIV/AIDS SERVICES

1. Describe the substance abuse treatment and/or outreach and pretreatment services to be expanded or enhanced in conjunction with HIV/AIDS services.
2. Describe how the proposed service or practice will be implemented.
3. Describe the process of offering and following up on rapid HIV testing of all clients and their injection and/or sexual partners, either during outreach, pretreatment or program enrollment. Describe your procedures to refer and confirm receipt of HIV treatment services for clients who test HIV positive.
4. Provide a realistic time line for the entire project period (chart or graph) showing key activities, milestones, and responsible staff. [Note: The time line should be part of the Project Narrative. It should not be placed in an appendix.] Provide evidence that the proposed expansion and/or enhancement will address the overall goals and objectives of the project within the 5-year grant period.
5. Clearly state the unduplicated number of individuals you propose to serve (annually and over the entire project period) with grant funds, including the types and numbers of services to be provided (to include HIV testing and counseling) and anticipated outcomes (See Appendix L). Describe how the target population will be identified, recruited, and retained.

Source: SAMHSA (2007, p. 24).

excerpt from an old grant funding opportunity, it is useful to examine it to see the type of information that might be requested in an Implementation Plan section. This excerpt comes from Targeted Capacity Expansion Program for Substance Abuse, Treatment and HIV/AIDS Services (Short Title: TCE/HIV), TI-07-004, released by SAMHSA in 2007 (SAMHSA, 2007). It counted for one-fourth of all points for the proposals. Each of the elements in Box 9.1 had to be in the grant proposal for the agency to receive full points for this section of the proposal.

When you examine a request for proposals closely, in the line-by-line, word-by-word way that a grantwriter must, you often see items or bullet points that seem repetitive or duplicative of other information you are asked to provide. Within the limits of the number of pages you have, the question becomes how to present a great amount of information succinctly and with as little repetition as possible, while still responding to every point of the RFP. One way to decrease your burden in the implementation plan section is to refer to your logic model.

USING YOUR LOGIC MODEL AS A FRAMEWORK FOR YOUR PROGRAM IMPLEMENTATION PLAN

While the information in the previous section from the federal government is very useful, you may be feeling that you're still not sure exactly what to do in order to be able to write a good implementation plan. Fortunately, the logic model that you created is extremely helpful in providing a framework to write your plan. Let's return to the logic model we used before to demonstrate how to create an implementation plan (see Figure 9.1). As you go through this section, write down each task that you feel needs to take place and how long

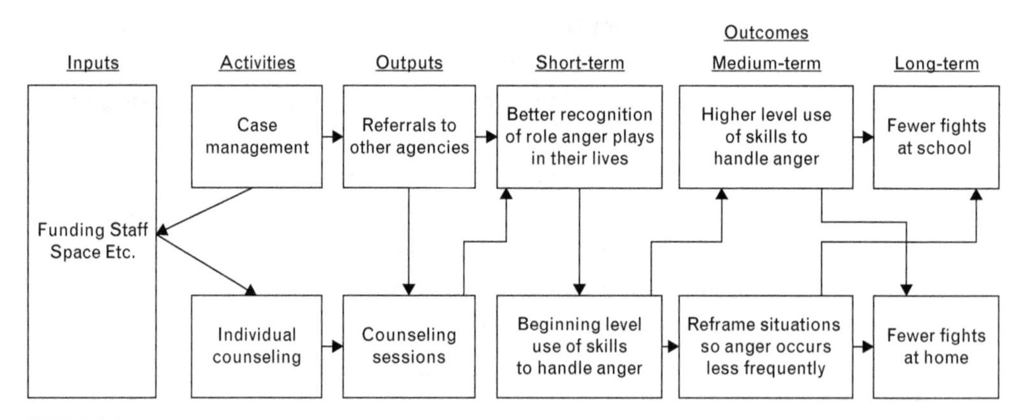

FIGURE 9.1: Example Logic Model (from chapter 7)

you believe it will take. At the end of this section, you can use this information to create a chart as requested in Step 4 of the example RFP.

INPUTS (RESOURCES)

The starting place to look in your logic model to help write the implementation plan is the first column: "Inputs." Listed under resources are these items: Funding, staff, and space. Your implementation plan *assumes* adequate funding, so there isn't really much else to think about here. Staff and space, however, need to be included in your implementation plan.

The first thing to consider is what type of staff members you need. If this is a new program, you'll have to consider support staff as well as clinical and managerial staff. The staffing plan has enormous implications for both how you implement the program and how you stay inside your budget. The logic model indicates that the activities of the program include both case management and individual counseling. The staff members you hire for this program thus have to be able do these activities. You may hire one person to do both or you may wish to have specialists. This decision will probably hinge on the number of clients you expect in both categories. You don't have to hire people for full-time jobs. The route you go here will of course have implications for your budget, which we cover in a later chapter, but it is vital to realize that there is a back-and-forth process between your implementation plan, your business model (what you do once the program is up and running), and your budget.

Let's make the assumption that the program we are developing that is depicted in Figure 9.1 will be started in a large school district but with only two high schools involved in a pilot test of the program. If it is successful, the program may be expanded. Let's make the further assumption that case management duties can be handled by a bachelor's level social worker (BSW) or person with similar qualifications. The individual counseling duties should be someone at a master's level, preferably a social worker (MSW) who is licensed at that level. Neither one needs to have prior experience in a school setting. It is probably the case that neither high school by itself will require a full-time person, but we believe that there will be a full-time caseload when looking at both schools together. We will thus assume that the employees will shuttle between the two schools. One of the job tasks for the MSW person will be to oversee the case manager and coordinate their efforts. This will require additional skills.

An alternative staffing solution would be to hire one person to do the program at each school. The advantage of this would be savings in transportation and space costs, as well as the one person being in complete charge of the program at his or her school, easing coordination issues. The disadvantage would be in needing to hire two licensed social workers, which would cost more than a BSW/BA-level worker and an MSW-level person. The case management tasks would be under-employing the MSW-level person.

Either approach is a possibility, but you need to make a choice and plan for it. New hires don't just walk in the door without having been recruited, interviewed, and made an offer. Set aside time in your plan to recruit applicants, interview them, decide whom to hire, and then bring them on board.

The funder wants to know who is responsible for the overall project. This will not need to be a full-time position by itself, and the duties will probably be assigned to someone already in place. In our example, this might be the Director of Student Success in the school district's central office. Because this program has activities in two different schools, it is probably not appropriate to assign the oversight duties to one principal. The task of managing this new project may take less than 10 percent of a person's time, or about four hours per week on average. You'll also need to think about the possible need for administrative support. Who will do this, and how many hours a week will it take? (See Box 9.2.)

Once you have covered all of the staff members needed in your implementation, you'll need to turn to the space required to run the program. Do the workers need office space? If so, where will it be located? Do clients need a place to sit while waiting to work with the staff members? Where will any physical files or supplies be located? What furniture will be needed, and when and how will it be procured?

ACTIVITIES

The second column in the logic model describes the activities that will take place. While the implementation plan doesn't say exactly how the program will be run on a daily basis, it needs to describe for the reader how you will get ready to perform the activities. For

BOX 9.2: WHAT IS A FULLTIME EQUIVALENT (FTE)?

You often hear of jobs being 1.0 FTE, or .5 FTE. What exactly does that mean? One FTE is the equivalent of one person working 40 hours per week. Most companies make the assumption that someone working full-time for a full year puts in about 2,080 hours (40 hours per week times 52 weeks in a year). Even allowing one week for vacation days and another week for sick days, that is 2,000 hours per year. So, when you say that you want to plan on 1 FTE, you need to consider how much work can be accomplished in that amount of time per week/month/year.

An important aspect about FTEs is that you shouldn't assume that 1 FTE is the same as hiring one person. If you hired four people for 10 hours a week, that would still be just 1 FTE. Given that frequently only full-time employees receive benefits, it can be less expensive to hire several part-time staff to make up 40 hours a week altogether than it is to hire one person full-time. Having part-time people receive pro-rated benefits does away with this cost advantage, but you probably get a more dedicated group of staff members if they feel they are treated well, despite being part-time.

example, once you have the qualified staff members hired, what will be necessary for them to be able to implement the program? Will they need additional training, as may be the case for specialized evidence-based programs? Will they need to be oriented to the host organization, such as a program taking place in a school or hospital? What supplies are required to implement the program? Put yourself in the shoes of the newly hired worker as you write this section. What do you imagine will be needed for you to do your work? Be sure to help that future hire, whom you probably don't even yet know, to have a good experience by laying the groundwork for success.

OUTPUTS

You may be surprised that you can use the outputs column from your logic model to assist you as you develop the implementation plan. You may have heard it said that if something is not documented, it didn't happen. As we write a grant and plan for a program's implementation, we must also plan how we will document both activities and outputs. Without doing so, it may appear that the program staff are spinning their wheels—doing lots of activities (such as case management referrals, or individual counseling sessions) but with nothing to show for their efforts.

As we have noted, doing activities and even having outputs occur is not the same thing as achieving change with clients. But, according to the program logic (as shown in the logic model), if the resources do not exist, the activities are not done, and the outputs do not occur, we should not expect the client outcomes to be realized. At this stage of the implementation plan, you need to decide how to keep track of the number of activities that take place and how you will document the outputs that result.

The logic model shows that case management activities led to referrals to other agencies. You should already be thinking about how to schedule the case management appointments and how to track completed appointments versus no-shows. Will you try to reschedule missed appointments? How will this be done? How about walk-ins? Will they be permitted? Once referrals are made, who will be responsible for making the appointment for the student? How will you track whether a referral is successfully completed?

Similar issues exist in terms of scheduling, meeting, and tracking students who receive individual counseling sessions. The main difference is that there are no outside actors to involve in terms of follow-up or outreach to outside actors.

The reason to answer these questions in your implementation plan is so that all systems can be put into place at the start of the program, without awkward gaps and realizations that the program has no method to track these matters. Funders will want to know how their money is being spent, and so it is necessary to pre-plan how to collect the data needed to develop reports of agency activities and outputs.

OUTCOMES

Columns 4, 5, and 6 relate to the desired and expected outcomes of the program. Measuring these is covered in the program evaluation chapter, as that topic is usually not included in the section on implementation. Program evaluation receives a separate section in the grant proposal. It is worth a considerable number of points in the total score of the proposal.

CREATING A TIMELINE (GANTT CHART)

As noted earlier, requests for proposals often require the inclusion of some type of chart showing the progression of the program's implementation and start-up. As one SAMHSA RFP indicated, "Provide a chart, graph, and/or table depicting a realistic timeline" (SAMHSA, 2013, p. 24). While not the only way to show the implementation of a program, a Gantt chart is easy to make and can be done in Excel, a software program that the vast majority of nonprofits already have. From there it can be pasted into the word-processing program being used to write the grant proposal.

According to the website gantt.com, Henry Gantt developed one version of the chart that now bears his name. At its most simple, a Gantt chart has a horizontal axis that represents time (the unit of measurement can vary from an hour to a month or longer) and along the vertical axis has the tasks associated with the implementation of the program or other project. The amount of time each task will take is indicated by a bar beginning at the task's start-date and ending at the end-date. The Gantt chart thus shows all tasks, how long each will take, when each task begins, when each task ends, which tasks overlap with which other tasks, and when the entire project will begin and end.

The chart developer begins by listing all the tasks to be completed and then determining when it will start, when it will end, and the number of days, months, or hours required. One can set this list out in an Excel spreadsheet where it looks similar to Table 9.1.

This information is then transferred to a chart using a process that is not difficult, resulting in a Gantt chart, which is shown as Figure 9.2. The exact process varies somewhat

TABLE 9.1: List of Tasks, Start Date, Duration, and Finish Date for Example Program

Task	Start Date	Duration (in Days)	Finish Date
Hire 2 MSW caseworkers	1-Jan	30	30-Jan
Select EB Program	1-Jan	15	15-Jan
Train caseworkers in EBP	1-Feb	15	7-Feb
Recruit clients	1-Feb	135	1-Jul
First Client Group	1-Mar	21	21-Mar
Second Client Group	1-Apr	21	21-Apr
Third Client Group	1-May	21	21-May
Follow Up First Client Group	21-May	15	7-Jun
Follow Up Second Client Group	21-Jun	15	7-Jul
Follow Up Third Client Group	21-Jul	15	7-Aug
Analyze Data	8-Aug	15	23-Aug
Write Report	24-Aug	7	1-Sep

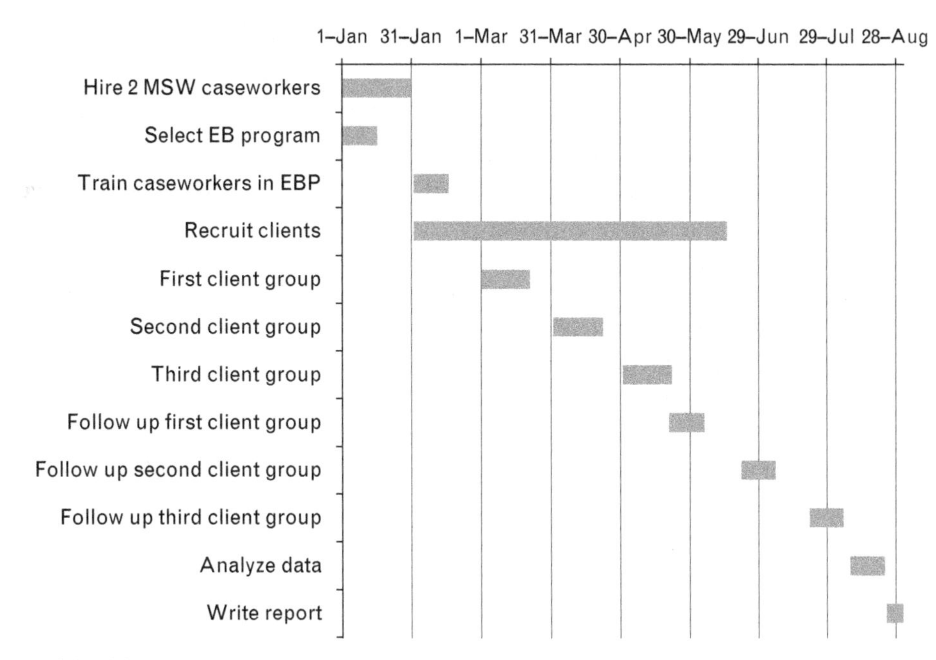

FIGURE 9.2: Excel Generated Gantt Chart Using Tasks List from Table 9.1

from one version of Excel to another, and a variety of videos exist on YouTube.com to show the step-by-step processes for different Excel versions. (One video that shows how to create a Basic Gantt chart in Excel was created by Doug H and is available online at http://www.youtube.com/watch?v=TjxL_hQn5w0.) Specialized project-management software can create more detailed Gantt charts that also link tasks, add constraints, include needed resources, and other refinements. For your grant proposal, a basic Excel-generated version is sufficient.

SUMMARY

The implementation plan section of a grant proposal is one where the grant writer must put on a "visionary" hat in order to see how things should unfold in the future. Just as when planning any big event, the ability to look at both the overall context and the specific details of the process is vital. One technique to see the context and details is to use a Gantt chart that is created in Excel. Program-planning software can make this chart more comprehensive, but a simple version is appropriate for grant applications.

Imagination and anticipating challenges are important aspects of the grantwriter's craft, and this is nowhere truer than in the implementation planning stage. The amount of effort you put into this section is considerable, but it will not be time that is wasted, even if you write more than what will fit into the page limits of the grant. The more you can lay out a clear plan for starting and running a program, the easier your organization will find it to put the ideas into place and serve clients, once the grant proposal is chosen for funding.

PRACTICE WHAT YOU'VE LEARNED

1. Find an RFP from an agency that requires that you create an implementation plan. Carefully delineate what the funder requires to be in this section. Do you understand everything that is requested? If not, find out the answers to your questions.
2. Look at the logic model you have developed for your grant proposal (you should develop a logic model even if your grant does not call for one, simply because it is so useful in creating the implementation plan and the evaluation). Begin the process of writing the implementation plan by recalling the funder requirements, reading the questions noted earlier, and looking at the resources needed (column 1). Walk through the questions earlier noted in this chapter and write a draft implementation plan. Include how you will monitor the number of activities and outputs.
3. Create a Gantt chart using Excel. Locate an appropriate tutorial online to walk you through the process. You can experiment using this type of chart for many things. For example, develop a Gantt chart for the following things:
   - implementing a program
   - writing a grant proposal
   - putting on a fundraising event

REFERENCES

H., Doug (2011). Create a Basic Gantt Chart. Retrieved from http://www.youtube.com/watch?v=TjxL_hQn5w0

Roper, A., Hall, T., & White, L. (2011). Best Practices for a Strong Implementation Plan. Office of Adolescent Health and Administration on Children, Youth and Families/ Family and Youth Services Bureau. Retrieved from www.hhs.gov/ash/oah/.../slides_implementationplanwebinar121310.pdf

Substance Abuse and Mental Health Services Administration (SAMHSA) (2007). Targeted Capacity Expansion Program for Substance Abuse, Treatment and HIV/AIDS Services (TI-07-004). Retrieved from http://www.samhsa.gov/Grants/2007/TI_07_004.pdf

Substance Abuse and Mental Health Services Administration (SAMHSA) (2013). Strategic Prevention Framework: Partnerships for Success (TI-13-004). Retrieved from http://www.samhsa.gov/Grants/2013/sp-13-004.pdf

www.gantt.com (2012). What is a Gantt chart? Retrieved from www.gantt.com

CHAPTER 10

BUDGETING

The subject of budgeting has been lightly touched on in previous chapters, but now we discuss the topic in greater detail. Up to this point you have been able to design the best intervention you can think of. This chapter forces you to consider whether the ideas you have can be put into effect with the amount of funding available. Even if your ideas are well conceived and grounded in the best of evidence-based practices, you need to match your request for resources to the ability of the funder (or funders) to provide them. Government and foundation program officers certainly want to see reasonable proposals and appropriate budgets to match.

WHAT IS BUDGETING FOR GRANTWRITING?

Budgeting for grantwriting, like budgeting for all things, links money with achieving outcomes. After all, the purpose of your grant application is to request funding to help make some problem go away (or achieve some positive goals). A budget in a grant proposal makes the case for how much money is needed to achieve the purposes of the funder through your organization's work.

The budget section is probably the most difficult to write for new grantwriters who have never put together a budget before or never had to adhere to one as a program manager. In fact, anyone writing a grant for a large sum of money will want to have the assistance of their organization's financial staff whether for government of foundation grant proposals. All budgets must be developed carefully and with great attention to detail. Once a grant is awarded, how the funds are used is closely watched. Nonprofits can get into serious legal trouble if the money is mismanaged.

WHAT GOES INTO THE BUDGET SECTION?

As with all aspects of grantwriting, you must refer to the request for proposals (RFP) documents from the funder to match what is required for a particular submission. For the purposes of this chapter, we'll look at an example federal RFP, the Competitive Abstinence Education Grant Program, HHS-2014-ACF-ACYF-AR-0827. We'll also discuss writing a budget for a submission to a foundation.

In almost all cases, government or foundation, there are two parts to the budget section. One is the actual dollar figures and the other is the justification or explication of the dollar figures. Readers want to know how much various elements of the proposal will cost and also why those elements are needed to achieve the purposes of the grant. The first part, showing the dollar figures, is usually done as a line-item budget although a functional budget approach may also be used. This is stated in the example RFP (Box 10.1).

Does this seem confusing? There is a lot of jargon here and references to forms and other things, like "object class categories," that you need to know about. Let's take these one by one and demystify the language. A project budget is a listing and categorization of expenditures to support the achievement of project outcomes. A line-item budget associates costs by "object class categories." Object class categories are categories of expenditures such as personnel, fringe benefits, travel, equipment, supplies, and so on. Thus a line-item budget will have one figure for personnel, another for fringe benefits, a third for travel, and so on. When totaled, the costs for all the object classes equal the total for the entire program/project.

Functional budgets take line-item budgets and provide additional detail, linking each line item to the achievement of a particular organizational function. If your organization has several different elements, the line items are listed separately for each one. To take a nonprofit human services program as an example, a residential substance abuse treatment program could have a functional budget that lists expenditures separately for the treatment program, food preparation, housing, and transportation. This allows managers to understand the costs of the agency in greater detail than if all these functions were put into one overall line-item budget. Another way to use a functional budget approach by an agency with several distinct programs can be illustrated by looking at an HIV/AIDS service agency. One program within the overall agency might be a prevention program;

BOX 10.1: THE PROJECT BUDGET AND BUDGET JUSTIFICATION

All applicants are required to submit a project budget and budget justification with their application. The project budget is entered on the Budget Information Standard Form, either SF-424A or SF-424C, according to the directions provided with the SFs. The budget justification consists of a budget narrative and a line item budget detail that includes detailed calculations for "object class categories" identified on the Budget Information Standard Form.

Project budget calculations must include estimation methods, quantities, unit costs, and other similar quantitative detail sufficient for the calculation to be duplicated (Administration on Children, Youth and Families, 2014, p. 22).

another a medical treatment program, and a third could be a behavioral treatment program. Each of these might be supported by a different funder, and it would be vital to keep the money from each funder segregated and clearly separated from the other programs. A functional budget could help the agency keep track of the separate streams of income and expenditures.

The dollar amounts you decide on for federal grants are entered into one of the standard forms used for this purpose. (Remember that foundations and other levels of government grants use different forms than these.) The SF-424A is to be used with applications that *do not* involve construction projects; the SF424C is to be used with applications that *do* involve construction of facilities. As you read this chapter, you may wish to view the online version of the SF-424A, which is available at http://www.acf.hhs.gov/sites/default/files/programs/css/sf_424a.pdf. A very useful aspect of the online version of the form is that it is an active spreadsheet, meaning that you put your figures into the form and it automatically calculates numbers for you, such as totals. See Figure 10.1 to look at SF-424A, Section A, Budget Summary. Extensive directions are provided on the actual SF-424A form for what should be entered in each line.

In Figure 10.1, which shows Section A of SF-424A, the first column is "Grant Program Function or Activity." If it is not required to use a functional budget, you would list the name of the program being proposed. The other columns are self-explanatory. If you are required to use a functional approach you would then need to list each function and the amounts for each function in the columns going across. Remember, this is the summary budget, not the line-item budget, which comes in Section B (see Figure 10.2).

Section B lists the Object Class Categories, or what we consider the "lines" of a line-item budget. If you are not using a functional budget approach, all costs per line go into Column 1. If you are using a functional budget methodology, then you must list each separate function across the top of the section and then enter the amounts for those functions in the appropriate row.

You need to include additional information in Sections C through F (none of which are shown here). Section C asks you to enter nonfederal sources of program support. These sources include state government, the applicant's own resources, and "other." Section D requests anticipated cash flow needs for the first year, by quarter, while Section E wants you to estimate the federal funding needed for each year of the grant period. In Section F, you must indicate the total direct costs (those costs that lead directly to client services) and indirect costs (those costs that are sometimes called "overhead" and can't be tied directly to the individual project in the proposal). This RFP states that organizations can only request indirect costs if they have an indirect cost rate agreement with the federal government and provide a copy of it with the application for funding.

Sometimes (although not in this solicitation), the funder requires the applying organization to provide matching funds, or a proportion of total project costs, so that the funder is not paying for everything in the proposal. The availability of these matching funds must be clearly stated as being already committed to the project. They cannot be "hypothetical" or something that will be found once the primary funding is assured.

BUDGET INFORMATION - Non-Construction Programs

OMB Approval No. 0348-0044

SECTION A - BUDGET SUMMARY						
Grant Program Function	Catalog of Federal Domestic Assistance	Estimated Unobligated Funds		New or Revised Budget		
or Activity (a)	Number (b)	Federal (c)	Non Federal (d)	Federal (e)	Non Federal (f)	Total (g)
1.		$	$	$	$	$ 0.00
2.						0.00
3.						0.00
4.						0.00
5. Totals		$ 0.00	$ 0.00	$ 0.00	$ 0.00	$ 0.00

FIGURE 10.1: SF-424A Section A Budget Summary

SECTION B - BUDGET CATEGORIES					
6. Object class categories	GRANT PROGRAM, FUNCTION OR ACTIVITY				Total
	(1)	(2)	(3)		(5)
a. Personnel	$	$	$	$	$ 0.00
b. Fringe benefits					0.00
c. Travel					0.00
d. Equipment					0.00
e. Supplies					0.00
f. Contractual					0.00
g. Construction					0.00
h. Other					0.00
i. Total direct charges (*sum of 6a–6h*)	0.00	0.00	0.00	0.00	0.00
j. Indirect charges					0.00
k. Totals (*sum of 6i and 6j*)	$ 0.00	$ 0.00	$ 0.00	$ 0.00	$ 0.00

FIGURE 10.2 SF-424A Section B, Budget Categories

HOW MUCH DO THINGS COST?

Two questions quickly arise when budgeting for a proposal: "What do I include in the budget?" and "How much do these things cost?" This is another time when it is useful to have a detailed logic model to turn to. It is also important to look at the section of your proposal related to program design and implementation, reviewing the section of the RFP on "allowable" and "nonallowable" costs.

Logic models have the desired outcomes listed on their right side. Outcomes are what we would like to achieve. On the left, however, is the information needed for the budget section. The key column is the one labeled "Activities." These are the aspects of the program that need a dollar figure attached to them. There are two aspects to this. First, what are the resources required for a particular activity? And second, how much do those resources cost? You probably have done much of the work for the first aspect if you have already completed the program implementation planning part of the proposal.

PERSONNEL

In the field of human services, usually the most costly aspect of any program is personnel. When it comes to knowing how much a particular position in the grant will cost, the primary considerations are the levels of education, training, experience, and skill required. For example, you may be proposing a program to reduce HIV/AIDS transmission among homeless people. What type of staff would you need and with what qualifications? When do they need to be hired? Who will have day-to-day managerial responsibility of the program and how much time will that take? How much of the top management's time should be allocated to oversee the background aspects of the project? Much of this is in the implementation section but if not, you need to flesh out that section before you can move forward. Besides salary costs, most agencies pay fringe benefits, such as vacation and sick leave, health insurance, and retirement benefits. Organizations are required to set aside money for other costs associated with staff as well. These costs need to be put into the submitted budget.

If you don't already have people employed at your agency in a similar job, you may not know how much to allocate for their salary. The Bureau of Labor Statistics (www.bls.gov) provides a quick overview of salaries for many occupations. As of May 2015, for example, the median salary for a social worker in the United States was $45,900. Salaries varied depending on the subfield. For mental health and substance abuse agencies, the median social worker salary was $42,170; for healthcare social workers, the median salary was $52,380 (Bureau of Labor Statistics, 2016, *Social Workers, Pay*).

If you want to know the salaries for social workers (or other type of worker) at the state or local level, you may try to use the Bureau of Labor Statistics website or use a third-party site. You can use a search engine such as Google to find out salary information. One site, www.salarybystate.org (2016) disaggregates information from the Department of Labor Statistics to show salary averages by state. The website www.salary.com (2016) shows salaries for jobs at the city level. Not only can you use these figures to plan your budget but also using these sources provides you with a good justification for the pay levels you set, in case the funder thinks you are paying too much.

TRAVEL

Nonpersonnel costs are needed, too. Nonlocal travel, such as to attend mandatory conferences or training, is expensive and must be included. This RFP states specifically that conference attendance is required for certain staff members and the travel costs must be included to receive full points when this section of the proposal is reviewed. The costs for this can be found on travel websites for airfare. Mandated conferences for federal grant recipients are nearly always in the home city of the agency making the grant. This means Washington, DC, for many agencies, but not all. The Centers for Disease Control and Prevention (CDC), for example, is based in Atlanta, Georgia, and CDC grantee meetings are held there. Hotels are arranged by the agency hosting the meeting, but the costs come from grantee funds. That is why they must be included in the proposed budget. Other costs of travel such as meals, local travel on-site, and so on, need to be estimated as well. Many agencies have their own standards for such things, and whatever is your current organizational policy should be followed. This can be sufficient justification for the proposal.

EQUIPMENT AND SUPPLIES

Both equipment and supplies are important categories. They are sometimes thought of as being the same thing, but are different. In general, equipment is more expensive (more than $5,000) and is considered as a nonexpendable asset. This means that it is expected to last more than one year. Supplies are lower in cost and expected to be "used up" within a year's time. Sometimes it may be difficult to determine which category is the correct one. A pad of paper is clearly a "supply" but what about a desktop computer or iPad costing only a few hundred dollars? We hope it will last several years, but it is well below the cost threshold given in the RFP. You may wish to seek advice from the contact person for the RFP to clear up any questions.

Once you have identified which equipment and supplies you will need, it is fairly easy (if tedious) to use an online supplier such as www.amazon.com to find their price. As you do this, be sure to keep track of the day you found the information and the supplier so that you have a record of the research for prices. Again, this will be your justification for the cost you cite in your budget.

CONTRACTUAL

Contractual costs include what the agency will pay for nonemployee staff and services. A common example is that of an outside program evaluator who is only employed for the project. The other common type of contractual situation is when the agency being funded intends to subcontract some of the work. If this is the case, you should be prepared to discuss the way the successful bidder will be chosen. The RFP does indicate that an "open and free competition" should be used for any contracts exceeding $150,000. A strong justification will need to be made when a no-competition contract is to be awarded. Of course, this may be easy enough to provide if there is only one provider of the required services, but it is important to be able to justify to an outside authority that the contracting process was in the best interest of the funder and the public.

OTHER

What should be included in the "other" category varies from one funder to another and so, as with everything else, needs to be double-checked against the language of the RFP. The Administration on Children, Youth and Families (2014) states other costs:

> include but are not limited to: consultant costs; local travel; insurance; food (when allowable); medical and dental costs (non-contractual); professional services costs (including audit charges); space and equipment rentals; printing and publication; computer use; training costs, such as tuition and stipends; staff development costs; and administrative costs (p. 25).

These other costs can be estimated and justified in the same ways as previously described costs. The key to any justification is that a reasonable person could see that you have made an effort to find a source of the needed skillset (in terms of personnel) or tangible good that is cost-effective. An important question to keep in mind is "If you were looking at the budget from the side of the funder, would you think it was a good way to spend your own money?"

REVIEW CRITERIA

For this federal RFP, the budget and justification section receives only four points out of 100. The reviewers clearly will have some chance to provide their opinions, but they do not have a large number of points to assign to this section. The two criteria that are listed for the reviewers to consider are: (1) "The applicant includes a detailed line-item budget of project costs and demonstrates how cost estimates were derived" and (2) "The applicant includes a proposed budget and budget justification that is feasible for the proposed approach; and is logical, reasonable and appropriate." Every point your application receives is perhaps the difference between receiving funds or not, so each section, no matter how few points are available for it, must be treated with great respect. If you have followed the steps and information provided in this chapter, you should be able to get full credit for this section of your proposal.

FOUNDATION PROPOSAL BUDGETS

The focus so far in this chapter has been almost entirely on the federal grant budgeting process. That is because if you can complete a budget for a federal proposal, then you are certainly ready to complete a foundation grant proposal budget. A search for a representative foundation to show the difference between foundations and federal government requirements brings up the conclusion that, unfortunately, there really is not a single "representative" foundation to use. Foundation requirements vary too much.

Foundations typically ask for a short letter of interest with limited information from the organization asking for funding. This information typically includes basic budget request information, but not to the extent that will be required if a full application is asked for by the foundation. This approach is common and points out an important difference between foundations and governmental approaches to grant making. Government-sponsored grant opportunities are much more rule-bound and attempt to treat each applicant within the same parameters in order to assure the public that no favoritism is shown. Foundations, on the other hand, may operate as they like in terms of their goals and processes, as long as they follow the law and their founding documents. This means that program officers working for foundations can help shape proposals, including their budgets, as the process unfolds. A foundation staff member can do a great deal more nurturing of an idea and shaping of a budget if a particular idea seems interesting.

One large family foundation in Dallas, Texas, for example, indicates on its website that it requires a letter of inquiry with limited budgetary information before deciding whether a full proposal will be requested:

> Please include a project line item budget including income and expenses. The table below is provided as a template that you are encouraged to use to submit your project budget data. Expense and revenue categories should be entered, along with corresponding dollar amounts for each. (Meadows Foundation, n.d.)

The actual budget form is a very simple line-item budget showing both expected expenses and revenues (see Figure 10.3).

BUDGETING AS AN ITERATIVE PROCESS

One of the reasons why budgeting for grant proposals can be difficult is that you may not know how much things should cost, and the amount of detail to keep straight can be mind-numbing. But another reason that budgeting may be difficult is that your wonderful programmatic ideas are too expensive and you need to reduce their scope (and cost). This is the iterative (repetitive) nature of grantwriting. It takes practice and time to be able to shape the project to the budget and the budget to the project. You may need to go back and forth several times before the two come together.

One agency I've worked with developed a foundation grant proposal costing $3.5 million per year for five years. While the foundation was interested and liked the idea very much, it was not going to fund the proposal for more than $1 million per year. Could the grantwriters cut so much of the budget out and still have their good idea come through? That was perhaps the most difficult challenge they had seen, but after two years of negotiation, the agency and the foundation came to an agreement over the scope of work and amount of support that would be committed. Now the project, which is considered a national model, is in operation. The amount of work that went into the original grant that had to be discarded was enormous, and the emotional pain of losing so much of the original idea was heart-wrenching. Still, the core of the effort was intact and has been

Expense.Categories	Fiscal.Year Amount
Total.Expenses	
Revenue.Categories*	**Amount**
Total.Revenues	

*When.listing.foundation.gifts,.please.list.each.gift.as.a.separate.revenue.line.

FIGURE 10.3: Example Foundation Budget Information Requested

implemented. The agency hopes to find other sources to expand the program to its original concept.

This process of reducing the scope and cost of a proposal takes place with governmental grants as well. It is important to be able to show exactly how your projected costs were calculated to attempt to defend your proposal from reductions. But in these times of scarcity, it may be unreasonable to expect to receive all of what you ask for. The lesson is to keep good records of why you say something will cost a certain amount, never pad your budget to make it easier to cut, but also be flexible with your ideas. Consider, if you must cut some parts of your project, what elements you might be able to implement first, as a separate idea, and then have the rest to phase in when more resources can be found.

SUMMARY

This chapter has introduced you to the vital area of budgeting for your grant proposal. We used a federal grant as an example and model because they generally require more detail in the application than do foundations. Still, we know that all proposal budgets require a great deal of work before the grant application is submitted. Foundation proposals are often less difficult than government grants because you usually submit an initial letter of intent without complete details.

As always, when you are preparing your budget, look to the RFP and other funder guidelines to determine what is allowed and what is not allowed in your budget. Use the following bullet points as a checklist when preparing a budget.

- When you read your budget and budget justification, do they come across as logical, reasonable, and appropriate?
- Is this a project you would fund if it were your money?
- Do all the elements that are described in earlier sections of the proposal seem necessary, and are they included in the budget?
- Are all the requirements of the RFP included in the budget? Is there anything that has been omitted, such as travel funds to attend grantee meetings, or other elements that have been included in the budget without being in the other parts of the proposal?
- Do the numbers add up correctly?
- If a line-item budget is required, are there lines that do not belong, or items in a line that shouldn't be there?
- If a functional budget is required, has it been completed properly?
- If you are asking for indirect costs, do you already have an indirect cost rate developed, such as through an agreement with the federal government or another foundation?
- Have you documented how you arrived at the prices and costs that you have included in the proposal's budget and laid that out in the budget justification section?

There can certainly be other issues to consider, but if you are able to confidently answer all these questions, then you will have done a good job with your budget and should be able to receive maximum points during the review process.

PRACTICE WHAT YOU'VE LEARNED

1. Using your logic model and the implementation plan that you have from earlier parts of the grantwriting process, develop a line-item budget including both personnel and nonpersonnel parts of the project. Be sure to keep a list of where you found the information for each element of your budget.
2. After you have completed the first exercise, be sure to have someone else look it over, using the criteria listed in the chapter summary. Ask for that person's honest feedback. Weigh it, and decide to alter your original budget or not. Have a good reason behind your decision.
3. Now, look at your budget and your plans for the proposal. Imagine you have been given only 85 percent of the amount you asked for. How could you reshape the proposal in order to meet the lower amount of funds available?

REFERENCES

Administration on Children, Youth and Families (ACYF) (2014). Competitive Abstinence Education Grant Program (HHS-2014-ACF-ACYF-AR-0827). Retrieved from www.acf.hhs.gov/grant/ope/foa/files/HHS-2014-ACF-ACYF-AR-0827_0.pdf

Bureau of Labor Statistics (2016). Occupational Outlook Handbook, Social workers, Pay. Retrieved from http://www.bls.gov/ooh/Community-and-Social-Service/Social-workers.htm#tab-5

Meadows Foundation (n.d.). Downloadable Grant Application. Retrieved from www.mfi.org/display.asp?link=YSBE5G

Salary.com (2016). Averages for Fort Worth, Texas, Social Workers. Retrieved from http://swz.salary.com/SalaryWizard/Social-Worker-BSW-Salary-Details-Fort%20Worth-TX.aspx?hdcbxbonuse=off&isshowpiechart=true&isshowjobchart=false&isshowsalarydetailcharts=false&isshownextsteps=false&isshowcompanyfct=false&isshowaboutyou=false

Salarybystate.org (2016). Social Worker Salary by State. Retrieved from http://salarybystate.org/healthcare/social-worker-salary-by-state

CHAPTER 11

AGENCY CAPACITY AND CAPABILITIES

This chapter addresses the section of a grant proposal that asks you to build a case for your organization's ability to deliver the program you propose. Because the topic has several subtopics, we cover a number of things that may seem at least slightly unconnected. The concept that ties them together, however, is that you are working to convince the funder that you are capable of using their resources in a way that will achieve the goals you set forth and that their funding is designed to achieve. After providing an overview of organizational capacity, the chapter describes the five main parts of this section of the grant proposal.

WHAT IS "ORGANIZATIONAL CAPACITY" OR "STAFF AND ORGANIZATIONAL EXPERIENCE?"

Your purpose in this section of the proposal is to convince the reviewers that your organization can do everything that you have indicated should be done. Box 11.1 is an example of the wording of a federal request for proposals (RFP) concerning organizational capacity. This comes from the Office for Victims of Crime, Office of Justice Programs, US Department of Justice, FY 2013 Services for Victims of Human Trafficking announcement, OVC-2013-3615. Naturally, it is highly specific in asking for past experience with providing services in trafficking. Other funding agencies will want you to show evidence relating to services in their field. Box 11.2 is from a different federal agency, and shows an analogous section from the Substance Abuse and Mental Health Services Administration Teen Court RFA. US Department of Human Services, Substance Abuse and Mental Health Services Administration (2012). These two sections from different departments show that the basic ideas of an "organizational capacity" section vary between different agencies but some requirements overlap.

BOX 11.1: ORGANIZATIONAL CAPACITY SECTION OF GRANT FOR OVC-2013-3615

Capabilities and Competencies: Applicants must state their experience with managing federal grants that support direct services to crime victims and document their administrative and financial capacity to manage federal grants.

Applicants must demonstrate a history of providing services on behalf of victims of human trafficking. Applicants applying for funding under the Comprehensive Services program area must describe their experience in providing services for all victims of human trafficking that may be identified within the targeted geographic area, including foreign national/U.S. citizen/legal permanent resident, adult/minor, male/female/transgender victims, and victims of sex and labor trafficking. Applicants under the Specialized Services program area must describe their experience in providing the proposed service for the population of trafficking victims identified in the proposal.

Each organization must also demonstrate that it has the expertise and organizational capacity to successfully undertake an initiative that involves significant collaboration with other agencies, including local, state, and federal law enforcement; victim service and faith-based organizations; local medical providers; and other community service providers to develop, expand, or enhance services to victims of severe forms of trafficking. Applicants must also describe how the program will be managed, the staffing structure, and include an organizational chart or other information describing the roles and responsibilities of key personnel.

Additionally, applicants must provide a list of personnel responsible for managing and implementing the major stages of the project; a specific plan for supervision of case managers supported under this project (if additional staff will be hired to complete the project, the applicant should attach a job description and the selection criteria for the position); and a description of the current and proposed professional staff members' unique qualifications that will enable them to fulfill their grant responsibilities.

Note: Key staff, including case managers, must have prior victim service experience or be under the direct supervision of a senior case manager or project director who has such experience (p. 17).

Source: US Department of Justice (2013). Retrieved from http://www.ojp.usdoj.gov/ovc/grants/pdftxt/FY13_Human_Trafficking.pdf

BOX 11.2: STAFF AND ORGANIZATIONAL EXPERIENCE SECTION FOR TEEN COURT GRANT (US DEPARTMENT OF HEALTH AND HUMAN SERVICES, SUBSTANCE ABUSE AND MENTAL HEALTH SERVICES ADMINISTRATION, RFA TI-12-004)

SECTION D: STAFF AND ORGANIZATIONAL EXPERIENCE (20 POINTS)

- Discuss the capability and experience of the applicant organization and other participating organizations with similar projects and populations. Demonstrate that the applicant organization and other participating organizations have linkages to the population(s) of focus and ties to grassroots/community-based organizations that are rooted in the culture(s) and language(s) of the population(s) of focus.
- Provide a complete list of staff positions for the project, including the Project Director and other key personnel, showing the role of each and their level of effort and qualifications.
- Discuss how key staff have demonstrated experience and are qualified to serve the population(s) of focus and are familiar with their culture(s) and language(s).
- Describe the resources available for the proposed project (e.g., facilities, equipment), and provide evidence that services will be provided in a location that is adequate, accessible, compliant with the Americans with Disabilities Act (ADA), and amenable to the population(s) of focus. If the ADA does not apply to your organization, please explain why.

Source: Retrieved from http://www.samhsa.gov/grants/2012/TI-12-004.pdf

These two examples from federal government RFPs point out both similarities and differences in what is required by different federal departments. As always, read the RFP you are applying for very carefully and use it to guide you. Typically, five elements are listed in the section regarding organizational capacity, although not every RFP will have all of them.

- Experience with this or similar programs and/or populations;
- Sufficient administrative and financial capacity and oversight to run a federal program;

- If a collaborative program, experience in leading or being a part of a collaborative system;
- Description of how program will be staffed and managed, including an organizational chart and/or information on the roles and responsibilities of key staff and the extent of time devoted by key staff members to this proposed project;
- Qualifications of current staff for working with this population and job descriptions of staff members.

We also cover one final topic that helps build a case for your organization's capacity to fulfill the grant requirements: Memoranda of understanding (MOUs) and letters of support.

EXPERIENCE WITH SIMILAR PROGRAMS AND/OR POPULATIONS

The key element here is to show that your organization has a history of working with people who are similar to those who will be assisted in the proposal you are writing or that you have provided the same type of proposed services to people who may be different from the population you propose to serve in this proposal. History can best be shown by laying out in chronological order, starting with the present, the program(s) and populations being discussed in your grant application. Nonprofits usually tend to specialize in one type of treatment or population, so this should not be too difficult.

If this is a new venture for the organization, going beyond what has been done in the past, grantwriters can sometimes feel that it is very difficult to get funding because it seems you must already have provided the services you are requesting resources for. One way around the issue of needing to "have a grant" before you can "get a grant" is to enlist and collaborate with well-qualified partners with a strong history of funding in this area. They become your "mentor" and thus they "sponsor" your organization to help you acquire related expertise. In a sense, the experienced organization is lending its experience to you so that both organizations may have a better chance of securing the funding that is needed.

Another approach is to try to receive smaller grants from other sources (such as foundations) that can get you started building a track record in this programmatic or client population area. This is likely to be a risky effort unless you can access foundation grant officers who support your organization and strongly believe in your capacity to create and implement new ideas.

It may be that you have experience with a population (e.g., people who are homeless), but not the precise services being proposed (e.g., substance abuse treatment services within a Housing First model). Or it may be that you have considerable experience with the type of programmatic approach (for example, Motivational Interviewing/ Stages of Change) but with a different population. You may have a history of working with chronic substance abusers but now you want to expand your agency's work to people with a high risk of contracting HIV/AIDS through unprotected sex. Ideally, you can show that your organization has a record with the type of service you are proposing and has worked with the population in question, so you can overcome any skepticism about your capacity. Because you may not be in this ideal situation, your job in this section of the proposal is to make the best possible case for your ability to implement and run the proposed program based on what you have done successfully in the past.

As part of indicating that your organization has the needed capacity, it is mandatory to already have highly qualified people working for you. You also need up-to-date and clear job descriptions for any new positions that will be created after the grant is awarded. Reviewers will look for evidence that your organization can hire and retain employees with the skills and experience needed to be successful in the proposed program. (See Box 11.3 for information regarding how to write any job descriptions you develop to be included in your grant proposal.)

HAVING SUFFICIENT ADMINISTRATIVE AND FINANCIAL CAPACITY TO RUN A FEDERAL PROGRAM

Receiving a grant for the first time is a wonderful infusion of funds. It is, however, also a demanding and somewhat overwhelming process. After the grant proposal has been awarded, but before the agency receives any money, the agency must meet a variety of standards that indicate a well-functioning organization. Of primary importance are internal controls and procedures that ensure that the funds will be well accounted for and used only for the purposes of the grant. Before you decide to write the proposal, be sure that you have adequate staff in the accounting department, with proper procedures in place to be able to handle high-level demands for tracking and segregating the funds. Be prepared to describe and articulate these procedures with a level of detail that communicates your commitment to responsibly administering grant funds whether from government or foundation sources.

COLLABORATIVE PROGRAM EXPERIENCE

In recent years, many funders have moved to a community or sector-wide approach to solving problems (Friedman et al., n.d.). These grantors believe that individual organizations, no matter how capable, cannot adequately address deep-seated issues such as homelessness, poverty, or substance abuse on their own. Agencies applying for grants in these cases must be prepared to work collaboratively with other organizations in a comprehensive effort to address the underlying issues.

Because this type of programmatic structure is challenging, funders want to know the likelihood of success in establishing and maintaining collaborations. Previous experience is helpful in learning how to use this approach to improve clients' lives. One of the most important lessons from studies on collaborations is that there must be clear processes for running the project. Four areas have been identified by Friedman and colleagues (n.d.) as particularly crucial for collaborations to flourish:

- **Distribution of funds within the collaboration**: How will resources within the collaboration be divided and reallocated, if needed?

BOX 11.3: EXAMPLE JOB DESCRIPTION: CASE MANAGER FOR TRIPLE-A PROJECT PROGRAM

Tasks: Case managers in the Triple-A Project program will work with clients directly; coordinate services with other organizations for individual clients; work cooperatively to make and follow-up on client referrals to other organizations; collect information for client services and program evaluation purposes; and maintain the highest levels of client confidentiality and ethical behavior as codified in existing regulations and the NASW Code of Ethics.

Methods Used: Case managers will use standard instruments provided by the Triple-A Project to assess clients for individual needs. Employing reference materials and personal knowledge of other agency and community resources, case managers match client strengths and needs with available resources, and, when necessary, advocate for different or additional resources to be created. Case managers will seek supervision from the Case Manager Supervisor to improve their handling of individual client outcomes as necessary.

Purpose and Responsibilities: The purpose of the Case Manager position is to improve conditions for clients. They interact directly with clients of the Triple-A Project to improve their functioning with the goal of attaining self-sufficiency. Case managers are responsible for daily positive interactions with clients and ensuring that clients are given access to programs and resources that lead them to reach self-sufficiency as quickly as possible. Records must be updated frequently and maintained electronically in a secure fashion as dictated by agency policy. Case managers must be able to work well in a team, taking direction from the Case Manager Supervisor.

Relationship to Other Jobs: Case managers report directly to the Case Manager Supervisor, who in turn reports to the Triple-A Project Program Director. All case managers are expected to cooperate with each other and other Agency personnel in the best interests of the organization and, ultimately, the clients. Working through the Program Director and Case Manager Supervisor, case managers are also expected to cooperate with requests made by the program evaluator to collect process and outcome evaluation information. Some direct interaction with the program evaluator may take place.

Qualifications: A Bachelor's degree in social work is the minimum level of education required. A Master's degree in social work is preferred. Licensure at the appropriate level is required within twelve months of being hired. Experience in a case manager position is beneficial.

- **Recruitment and enrollment of clients:** How will clients be brought into the collaborative system? Who gets "credit" for helping achieve collaboration goals?
- **Access to client records**: How will client records be accessible appropriately to all involved organizations? Who is in charge of client data management?
- **Collecting evaluation data**: How will data collection and entry regarding project outcomes be handled?

If you are proposing a collaboration to address a problem in your community, your experience in handling these topics is of considerable interest to the potential funder. Evidence that you have encountered and overcome challenges should be highlighted in the organizational capacity section of the grant proposal. If you do not have any experience with collaborations, it may be better to acquire some before taking the lead role in such a proposal.

DESCRIPTION OF PROGRAM STAFFING AND MANAGEMENT, INCLUDING ORGANIZATIONAL CHART

As noted in the excerpts from two RFPs (Boxes 11.1 and 11.2), you are expected to be able to describe how the proposed program will be staffed and managed, and to include an organizational chart to demonstrate organizational capacity. An organizational chart is a snapshot of your agency's structure that illustrates staffing and reporting responsibilities. A lot of information can be packed into a small amount of space in your proposal by using a chart.

In very small organizations, there is little structure and the staff members may be called on to do any and all tasks. The founder of the organization, for example, may write grants, develop marketing plans, solicit donors, work with clients, get the mail, buy supplies, answer the phone, provide her own IT support, and everything else. This is not really a sustainable approach to running an organization, however. Once sufficient funds come in, one or more other people are hired to do some of these tasks.

Organizations that are capable of handling the demands of a federal grant, or any but the smallest foundation grants, will have more structure than just a CEO. An organizational chart allows an outside reviewer to look at how logically the agency is put together, how many people are in the organization, and how the new initiative being proposed fits into the existing situation.

Organizational charts also show who reports to whom, which is important to illustrate the chain of command and who is directly supervising the program and its workers. Just as with a logic model, a well-done organizational chart can convey a great deal of information to someone who understands their power. Charts like this can be made easily using Microsoft Word and Excel. Here is a link for a YouTube video showing the process of developing an organizational chart in MS Word, once you know what positions and individuals need to be shown in the chart and where they belong (http://www.youtube.com/watch?v=mDZrBxzfmOg).

The following three tips from the literature will help you develop an organizational chart for your grant proposal, if one does not already exist:

- **Typically, organizational charts are hierarchical**, with the chief executive officer (CEO) at the top of the pyramid. The row below the CEO shows the positions that report directly to the CEO, a relationship that is denoted by a solid line. The next row shows the people who report to the CEO's direct reports, and so on. Organizational charts can show more or less detail, depending on the size of the organization. The International Red Cross, for example, would have much less detail in its organizational chart than would a local domestic violence shelter. There are other organizational chart styles that show different types of organizations, but the hierarchical model fits almost all human service agencies.
- **Know the difference between line and staff positions and show them appropriately**. A line position is directly working to accomplish program goals, while a staff position is supporting program goals indirectly. For example, a caseworker and a casework supervisor would be line positions whereas an evaluator, a grantwriter, a human resource manager, or an accountant would be in a staff position. This difference is important to funders because they generally want to see their resources going to support line positions and their supervisors, even while knowing that staff positions are essential for organizations to function.
- **Size matters**. The boxes for people higher in the organization are typically larger than are the boxes for people lower in the hierarchy. People who are considered peers in the organization have boxes of the same size.

To draw an organizational chart for a program that does not yet exist (as is appropriate for a grant application), one could follow these steps using pen and paper, or software such as Microsoft Word (see Figure 11.1 for an example made using SmartArt in Microsoft Word).

1. Begin by drawing a box for the program director.
2. On a row below that box, draw a smaller box for each of the positions that will report to the program director (or, put another way, for each person whom the director will supervise).
3. Draw a horizontal line between each person at this level.
4. Draw a vertical line from the horizontal line to the program director box.
5. Repeat for each level in the hierarchy, showing which positions report to which positions already on the chart. Each level's boxes should be smaller than the ones above.
6. To position this program in the larger agency, draw a vertical line from the program director. At the top of this line, place a box that is the position that the program director reports to. This new box should be larger than the program director's box.

In Figure 11.1, we have the proposed program (the Triple A Project) illustrated. Five line positions and one staff position will be created. The program director will directly supervise two positions, the case manager supervisor and the client recruitment and follow-up specialist.

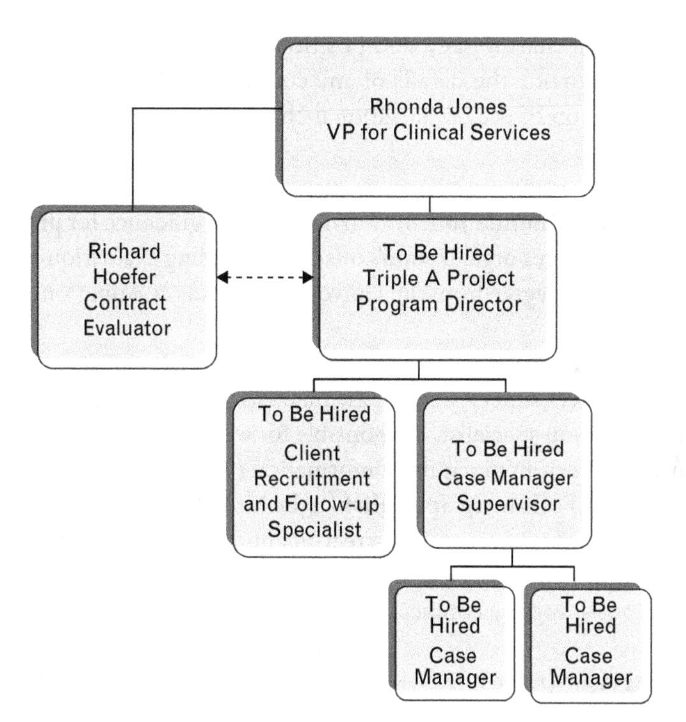

FIGURE 11.1 Example Organizational Chart Created in MS Word, Using SmartArt

The case manager supervisor has two case managers to supervise. None of these people have yet been hired. The organization will need to create job descriptions for these positions so that it is clear what their jobs entail. The Triple A Project program director reports to an existing position, the vice president for clinical services, Rhonda Jones. Most funders' RFPs would require you to provide her qualifications to be the overseer of the new program.

In addition, a contract program evaluator position is created, and Dr. Richard Hoefer will be hired. This is a staff position, as he will not work directly to accomplish the program goals but will support the achievement of program goals by monitoring and evaluating the program. This position works with the program director (as shown by the dotted line) but reports to and is responsible to the vice president for clinical services (shown by the solid line). His resume, showing his qualifications to be the program's evaluator, needs to be included with the proposal. (If the program has not yet located a program evaluator, this position still needs to be on the organizational chart and needs to have a job description developed and included with the other materials.) Also required is an MOU indicating that he commits to providing the evaluation services.

Most agencies have an organizational chart already developed for the larger organization. While that is an important piece of "boilerplate" information that should be on the hard drive of every grantwriter, it is vital to situate the new program within the existing structure. In fact, getting the new program integrated into operations adequately may require changes to the existing reporting structure, so the grantwriter needs to think through the implications of creating a new program carefully.

Another aspect of the description of the program's staffing is the listing of key staff, what they will do, and the extent of their time devoted to this project. In this case, we can look at the organizational chart for key staff positions and create a list quickly. Other agency documents should provide the details of any current workers' qualifications and background. Starting at the top of the organizational chart (Figure 11.1), we briefly discuss each position.

VP for Clinical Services (Rhonda Jones): Provides overall guidance for program and its integration into the larger organization's mission, including evaluation (5% time). Ms. Jones has an MSSW degree from the University of Texas at Arlington, with 10 years' experience.

Triple A Project Program Director (to be hired): Supervises all aspects of Triple A Project, including direct supervision of case manager supervisor and client recruitment and follow-up specialist. Responsible for working with program evaluator to ensure collection of required information (100%).

Client Recruitment and Follow-up Specialist (to be hired): Works with other organizations and community resources to recruit appropriate potential clients into the program and follows up with them to promote retention during the program. Is key link in having post-program contact with clients in order to collect evaluation data (100%).

Case Manager Supervisor (to be hired). Supervises two case managers and provides case management services on an as-needed basis. Reports to Triple A Project program director. Ensures grant-related reporting and data collection tasks are completed by case managers (100%).

Case Managers (two positions) (to be hired): Work directly with clients to assess and assist them in achieving their treatment goals (100%).

Program Evaluator (Dr. Richard Hoefer): Contractor responsible for developing and implementing evaluation plan to meet needs and requirements of organization and funder (10%). Dr. Hoefer has a PhD from the University of Michigan in social work and political science and has been lead evaluator on eight other SAMHSA grants.

QUALIFICATIONS OF CURRENT STAFF AND JOB DESCRIPTIONS

Getting a grant that creates new positions can be beneficial for organizations that have few other promotion opportunities. Current staff members who could be promoted to work in the new program if it is funded have several advantages over new hires. First, they are already employees in the organization and have vital skills in their current position. Thus, they may only need to be assigned the new duties. As an example, someone who is an experienced case manager now may be able to be promoted to case manager supervisor. Someone who works well with clients in a staff position, such as an intake coordinator, might be considered appropriate for the client recruitment and follow-up specialist position.

No matter who is assigned to the positions, it is important to the funder that they are qualified. Qualifications can include educational achievement or experience from work or volunteering. The listing of key staff (above) has a one-sentence statement of qualifications for the two known people (Rhonda Jones and Richard Hoefer) who will be involved with the program. These statements should be substantiated with resumes in an appendix.

We have noted several times the need to include job descriptions for proposed positions that are new to the organization. Job descriptions are an important part of running a human services organization. They are particularly important when needed in a grant proposal to show the potential funder that you clearly understand the tasks and roles of the positions in your program.

A well-written job description is helpful to your organization as well. It helps ensure that you will have qualified and competent employees implementing the program once the program is fully staffed. Job descriptions help attract people with the characteristics and background you want for the job, including education level and previous experience. It assists potential applicants by alerting them to job tasks and expectations before they apply for a position, thus reducing the number of inappropriate applications. You may be able to use currently existing job descriptions from your organization, borrow from another organization to edit, or create one from the ground floor. As long as what you turn in is appropriate for your organization and the program you are proposing, you will have an appropriate job description. This section describes the elements needed in an effective job description and shows how they are used (see Box 11.3 for the example job description).

A job description is useful when writing an advertisement for the position, but the two are not the same. A job announcement will typically include a salary or salary range (although not all advertisements are specific about salaries for jobs). Advertisements include how to contact the agency and what the requirements are to apply for the job, such as a transcript showing graduation and/or the names and contact information for references. This type of information is not included in a job description. According to the Small Business Administration (SBA, n.d.) the following elements should be included in every job description:

- Individual tasks involved
- The methods used to complete the tasks
- The purpose and responsibilities of the job
- The relationship of the job to other jobs
- Qualifications needed for the job

Following the SBA's guidelines, a job description of a case manager for the program depicted in Figure 11.1 might look like what is in Box 11.3. Naturally, there is considerable leeway in how a job description is written. Looking at other examples can be useful. One such example, for a Case Manager 1 position, which is more detailed than the job description in Box 11.3, is available from the Hamilton Care Center, a treatment facility in Indiana, at this website: http://www.hamiltoncenter.org/hr/JobDesc/Generic/CaseManager.html.

The Hamilton Care Center's job description has these headings in its position summary:

- Essential duties and responsibilities
- Minimum qualifications/requirements

- Certificates, licenses, registrations
- Physical demands
- Work environment, and
- Conditions of employment

The job descriptions for any organization need to be updated every now and then. Positions evolve, and the skills needed to do them well change over time as well.

MEMORANDA OF UNDERSTANDING AND LETTERS OF SUPPORT

It is quite true that you need to paint a picture for the funder of how capable your organization is and how well qualified your staff members are. In reality, though, no agency can fully stand by itself. Most funders want to see that the applicant can reach out to work with other agencies and service providers to have a greater impact on the proposed client population. Many RFPs now give extra points or require the applicant organization to engage in collaborative efforts in order to receive funding. Descriptions of collaborations are not enough, however, and must be backed up with MOUs. Even when a proposal does not involve a formal collaboration, obtaining clear support from other agencies and individuals in your community is an important element of your agency's capacity to be successful.

Memoranda of understanding are formal agreements between the agency writing the grant (often called the lead agency) and other entities. As an example, we have briefly mentioned the program evaluator, who as a contracted worker, must agree to provide services for this grant. This can be done with an MOU. Memoranda of understanding also come from other organizations that, for instance, agree to provide referrals of clients to the lead agency, or to provide services to clients if the lead agency refers clients to them, and MOUs can also be used to help ensure that the project is successful. An example of the need for an MOU is when a lead agency is seeking housing for people who are homeless as it provides them with mental health services. As a mental health provider, the lead agency probably does not have its own supply of housing units, so it must find another agency that does and will agree to set aside a certain number of its housing units for the lead agency. In this example, a formal agreement (an MOU) to provide services from the housing agency to the mental health agency will be required by a funder before any grant is given.

A letter of support is a way for a grant-seeking organization to indicate it has support from the community for the ideas in the grant proposal. Letters of support are not formal agreements in the same way that MOUs are—they simply are expressions of agreement that the goals and purposes of the grant are beneficial and needed for the community at large. Grantwriters will frequently orchestrate these letters, either directly or through the work of the organization's executive director, who can call on her peers at other organizations for help.

Often, the request for a letter of support will include possible language to use. This reduces the time for the writer of the letter of support, who can just cut and paste the material onto a piece of agency letterhead, scan and send back to the grant-seeking organization. There are disadvantages to this approach as well, though. First, the person signing the letter of support probably has only a very slight idea what is in the grant proposal and cannot be expected to be 100% behind the effort. Second, if several "authors" of letters of support all say the same thing because they used the same "ghost-written" materials, the reviewers may discount all the letters.

It is important to be aware that the act of asking for (and receiving) a letter of support for your grant application informally obligates you to return the favor when the other agency is writing a grant. Thus, it is useful to know how to write a letter, whether you are writing a draft letter to send to other agencies for your program or are writing to support another organization (Box 11.4). Here are some basic tips.

- The salutation often says something as generic as "To Whom It May Concern," but it is far better to have an actual name in the salutation.
- The first paragraph should say you support the grant proposal being written by Agency XYZ. If you have the exact name and RFP this should be included.
- The second paragraph should indicate that you are familiar with the need being addressed by the grant and that it is a true problem in your community/state/area. A few summary statistics or indicators are useful here.
- The third paragraph can indicate that the program proposed by Agency XYZ is a good approach to the problem, because it is innovative, evidence-based, collaborative, or other adjective that fits the goals of the RFP. Being specific about the details of the program theory or structure is helpful as it shows a real familiarity with the grant.
- The final paragraph should repeat that you support the proposal and that you can be contacted for further information. You will almost never be contacted if the application is for a federal or state grant, but a foundation grant officer might actually do so. It is thus in your interest (and the interest of the grant-seeking agency) that you sincerely know about and support the details of the application.

SUMMARY

This chapter has covered a lot of ground, all of which is designed to help you communicate the capacity and capabilities of your agency. Given everything that has been written previously in your application, you must show that you actually have the ability to make the proposed program a reality, should you be awarded funding.

Some of the aspects of creating a capacity statement that is convincing are to highlight agency experience with similar programs and/or populations; demonstrate that existing and new agency administrative structures and financial capacity are well integrated and set up to accomplish program outcomes; show a history of collaborative efforts (if this is a collaborative project); discuss qualifications of current staff members to implement and run the program, and describe the positions to be funded in enough detail to allow reviewers to ascertain whether you grasp what is needed for the program to be successful. Information

BOX 11.4: SAMPLE LETTER OF SUPPORT

Dear Mr. Grant O. Ficer:

I write to express my strong support for the proposal written by Agency XYZ to provide comprehensive, collaborative and client-centered services to the homeless in their Triple A Program proposal in response to RFP 2014-12-111, for Homelessness Reduction Services.

Homelessness continues to be a problem in Ourtown, despite strenuous efforts to reduce the numbers. Last year, Ourtown's homeless census counted a 5% increase overall compared with the year before, with an even greater increase among children aged 3–15 (32% increase). Families in Ourtown continue to struggle to make ends meet. The negative impacts of high unemployment (a state high of 24%), home loan defaults (up 18% over 3 years ago) and decreasing state and federal funding for basic services have combined to cause the problem of homelessness for many of our residents.

The program being proposed by Agency XYZ provides a path out of homelessness for participants. The Triple A program ("Assessment and Applications for Assistance" Program) is a low-cost yet innovative approach and will assist homeless people (particularly targeting families with children between the ages of 3 and 15) by ensuring that they have knowledge of all programs that could help. After a careful assessment of their needs, clients will be guided through a series of referrals and applications for assistance to meet their needs. The cost of the program is modest but the impact will be considerable. The strong evaluation component of the proposal will ensure that the program is able to serve as an evidence-based model, should it be as successful as we anticipate it will be.

As President of Ourtown's Coalition Against Homelessness, and on behalf of the 12 agencies in the Coalition, I strongly support this application. If you have any questions, please do not hesitate to contact me.

Sincerely,
Veranda P. Atio, MSSW
President, Ourtown's Coalition Against Homelessness

on MOUs and letters of support is also in this chapter so you can present a fully finished picture of your organization's proposal. If you provide information in ways similar to what has been described in this chapter, you will be getting close to the stage where the letters you receive from funders will contain the word "Congratulations!"

PRACTICE WHAT YOU'VE LEARNED

1. Look over the previous chapters of this book and the sections you have already developed, particularly the solution to the problem (your new program) and your implementation plan. List your agency's experiences with client populations, dividing your list into intervention areas (such as substance abuse, mental health disorders, prison re-entry) and demographic characteristics (racial, ethnic, sexual orientation, age, gender, and so on.) Using these entries, think about how you can show your capability to implement and run your proposed program. Do some of your current staff members have experience being in or managing a similar program, or have they worked with the population you seek to assist? Look at your organization with a critical eye—are you really able to do this project on your own? If so, explain why. If not, examine your community's other agencies to determine which of them can help you reach the capacity to do so. Reach out to them to ascertain their interest in partnering with you on this proposal. It takes courage to admit that your organization may not be up to the task by itself, but it is better to determine this before submission than to move forward with a project that is beyond your organization's current capacity.
2. Beyond the programmatic knowledge and skills embedded in your agency, what elements of your administrative structure will allow you to manage and be responsible for the financial controls required of large grant recipients?
3. If needed for this proposal, review and write up your experience with collaborative efforts. What went well? What didn't go well? How would you do a better job overcoming the challenges this time?
4. Reviewing the program plan and budget, as well as any existing organizational chart, develop a new organization chart that includes all staff positions that are part of the proposal. Follow the guidelines provided in the chapter. Be sure to include all staff positions that are included in the budget for the proposal, including staff members who are already employed but will be a part of the new program.
5. Look carefully at the information you have on current employees. Which ones fit "as is" into the new program? Describe what they will do if the proposal is funded. Be sure to have their current resumes and be able to pinpoint their qualifications for working in the new program.
6. For new positions, develop job descriptions that provide reviewers enough detail to determine whether these jobs are necessary to accomplish program outcomes. Use the ideas in the chapter to aid you in this process.
7. If your proposal requires the assistance of one or more other agencies, list which agencies you will approach. Note which services will be needed from other agencies, and determine which organization(s) you will approach to obtain the resources. Come up with a list of organizations and their CEOs for quick reference. Draft an MOU for their consideration.
8. Draft a letter of support for your program that you can send to leaders of other organizations who may be willing to lend a hand to your proposal. Follow the suggestions provided.

REFERENCES

Friedman, J., Sutor, C., Warfield, T., Gallant, L., Gettings, R., East, B., Glover, R., Kashen, K., Powers, S., Hollan, J., & Van Lare, B. (n.d.). Opportunities for Collaboration Across Human Services Programs. Washington, DC: American Public Human Services Association. Retrieved from https://web.archive.org/web/20090106000350/http:/www.financeproject.org/Publications/EBO_collaborationprograms.pdf

Hamilton Care Center (n.d.). Case Manager 1 Job Description. Retrieved from http://www.hamiltoncenter.org/hr/JobDesc/Generic/CaseManager.html

Small Business Administration (SBA) (n.d.). Writing Effective Job Descriptions. Retrieved from http://www.sba.gov/content/writing-effective-job-descriptions

US Department of Human Services, Substance Abuse and Mental Health Services Administration (2012). Teen Court Grant, RFA TR-12-004. Retrieved from http://www.samhsa.gov/grants/2012/TI-12-004.pdf

US Department of Justice, Office of Justice Programs, Office for Victims of Crime (2013). OVC Fiscal Year (FY) 2013 Services for Victims of Human Trafficking. Retrieved from http://www.ojp.usdoj.gov/ovc/grants/pdftxt/FY13_Human_Trafficking.pdf

CHAPTER 12

FINAL DETAILS

In this chapter, we cover a few of the final details that haven't fit nicely into other chapters. We cover the topics of sustainability, cultural competence, developing a title and writing the abstract, submission of the grant proposal, and how the review process works. We then close with a short postscript.

SUSTAINABILITY

A comparatively recent aspect of writing grants is the emphasis on "sustainability" or the ability to keep the program going after the grant period ends. No single funder wants to or can be expected to be the financial support for a program indefinitely. This section is a way for you to show you're thinking 3–5 years down the road when the initial funding is completed. What will you have in place at that time to keep the program going? (See Boxes 12.1 and 12.2 for examples of what is being asked for in terms of sustainability in federal requests for applications [RFAs].)

In order to sustain your proposed program it is vital to integrate it into the "everyday" of your organization. When you developed an organizational chart that included your new program, you made the leap from a project that did not exist at all to one that fit in well with

BOX 12.1: EXAMPLE PROJECT SUSTAINABILITY PLAN: TEEN DRUG COURT. RFATI-12-004.

Describe your plan to continue the project after the funding period ends. Also, describe how program continuity will be maintained when there is a change in the operational environment (e.g., staff turnover, change in project leadership) to ensure stability over time.

Source: US Department of Health and Human Services, Administration for Children, Youth and Families (2012).

BOX 12.2: EXAMPLE PROJECT SUSTAINABILITY PLAN: TRAFFICKING WITHIN THE CHILD WELFARE POPULATION

Applicants must propose a plan for project sustainability after the period of federal funding ends. Grantees are expected to sustain key elements of their grant projects, e.g., strategies or services and interventions, which have been effective in improving practices and those that have led to improved outcomes for children and families.

Describe the approach to project sustainment that will be most effective and feasible. Describe the key individuals and/or organizations whose support will be required in order to sustain program activities. Describe the types of alternative support that will be required to sustain the planned program. If the proposed project involves key project partners, describe how their cooperation and/or collaboration will be maintained after the end of federal funding.

Source: US Department of Health and Human Services (2014, p. 21).

the rest of the agency. A successful sustainability plan keeps you from having to erase that section of the organization chart.

In order to be sustainable, the program should be effective in improving the lives of clients. If it doesn't do that, there is no reason to keep it going. This is a good reason to have an excellent evaluation of the program (both process and outcome)—to ensure that you only hang on to programs that work for the benefit of your clients. Assuming that you have a solid, evidence-based program in your proposal, you should find that the program works. And, if the program works, it should be possible to keep it going, even in the Age of Scarcity.

The Office of Community Services (OCS) provides a very useful website to help readers develop sustainability plans (www.acf.hhs.gov/programs/ocs/resource/creating-your-sustainability-plan). The key questions the OCS believes an organization should ask include:

- Do the organization's mission and goals match the community's needs?
- What is the purpose of the program?
- What is the best way to contribute to the social service provider network? and
- Which other organizations should we collaborate with?

Once these questions are answered then the organization doing the planning needs to determine what their resource needs are and begin asking for that amount of resources. One of the key strategies for sustaining a successful program involves creating and maintaining partnerships and collaborations where resources are shared.

If you consider all the ways that nonprofit organizations bring in resources, you can begin to create a sustainability plan by putting some or all of these techniques to use. Here are a few to get your thoughts moving:

- Funding the program with a special social enterprise project, such as having clients produce goods that they are paid for but are sold to others for a profit;
- Dedicating revenue from the sale of information products (Hoefer, 2012) to support the program;
- Having a special event or series of events that are linked to the program to raise awareness and funding;
- Finding one or more corporate donors willing to make a long-term commitment to the program in exchange for being publically acknowledged;
- Finding major donors who will contribute annually or as bequests to help set up an endowment fund particularly for that program;
- Using volunteers instead of staff members to run programs.

It is unlikely that using only one of these sources would guarantee the sustainability of a program, but in combination with other ideas and making the program a "permanent" part of the organization's service offerings, these ideas can lead to a very well-defined sustainability plan.

CULTURAL COMPETENCE

Funders are well aware of the racial, ethnic, cultural, and linguistic differences that exist within the United States. There is a vibrancy that comes when human service organizations seeking resources connect with and support all of the cultures that are represented among staff members and clients. But this does not always happen in a meaningful way. Too often in the past the majority culture has had a near monopoly on receiving government and foundation grants with little or not enough understanding of their clients' cultures. In an effort to overcome this history, some grants now require organizations to include a section showing their cultural competence. Even when this is not a required section, you are always correct to ensure that your proposal is seeking to create or expand culturally competent services and programs.

To assist in planning services that are accessible by all consumers, the National Culturally and Linguistically Appropriate Services (CLAS) Standards in Health and Health Services have been developed. (These standards are widely applicable, not just in the health arena.) The intent of the standards is to "advance health equity, improve quality, and help eliminate health care disparities by establishing a blueprint for health and health care organizations" to implement the other standards (Office of Minority Health, 2013). The Principal Standard of the CLAS is: "Provide effective, equitable, understandable and respectful quality care and services that are responsive to diverse cultural health beliefs and practices, preferred languages, health literacy and other communication needs" (Office of Minority Health, 2013). The other fourteen standards deal with the areas of:

- Governance, leadership and workforce
- Communication and language assistance and
- Engagement, continuous improvement and accountability (Office of Minority Health, 2013).

Incorporating these standards (as appropriate) into your grant proposal will assist you in developing a proposal that shows an interest in meeting the need for culturally and linguistically appropriate services. As the grantwriter, you should carefully review all of the request for proposals (RFP), looking for the funder's desire to support organizations that are culturally competent and bring this to the attention of agency executives and staff. This information may not be highlighted in the most obvious places. For example, in one seemingly obscure sentence in a section titled "Objective Review and Results," the Administration on Children, Youth and Family (2013, p. 140) states: "A record of demonstrated effectiveness in providing services that are culturally and linguistically relevant to the populations being served" is necessary for receiving funds.

DEVELOPING THE PROPOSAL TITLE AND WRITING THE SUMMARY/ABSTRACT

Most RFPs have nothing to say regarding your project's title other than that you should have one. The title is your reader's first impression of what the proposal is about and should be chosen carefully. The title and the abstract will be public knowledge if your proposal is funded as a government grant, so they must communicate to both lay and technical audiences. Foundations usually publicize project titles and abstracts as well. A proposal title can be thought of as the "bumper-sticker" version of the entire proposal, or in this age of social media, the Twitter version of your proposal. A few tips for titling your proposal include:

Titles set the tone for your proposal and, along with the abstract, are often used to categorize your project, placing it in groups with similar themes for the convenience of reviewers. Be sure that your title is accurate and informative.

- Titles need to be specific and show that the project is significant.
- Including information on how funding will be used provides a helpful image for reviewers.
- Do not include a question in your title.
- Titles frequently have a limit to the number of characters allowed. Be prepared to draft several different versions to ensure you meet the requirements.

Proposal titles often include the name of the program that will be funded. Program titles are a chance to be creative and to develop a "catchy" brand. Here are some titles of programs that were funded in the 2011 grant competition for Strengthening Predominantly Black Institutions Program:

- Students Achieving Success in Engineering Technology (SASET)
- The Student Teacher Advisement and Retention (STAR) Center
- Project RISE (Realizing and Inspiring Success through Education)
- The Minority Male (M2) Leadership Initiative

A bad title probably never lost grant funding by itself and a great title won't make up for a proposal that is shoddy in other ways. Still, putting time into developing a program name that connects with potential clients, staff members, and funders is time well spent. Even as the grantwriter, you can look for ways to assist in client and staff recruitment by having a name for the project that is inspiring.

A summary (or abstract) of your proposal is a test of your writing skills. Most grant proposals are quite long, and you have spent a great deal of time on all of the sections, whether this is for a government grant or a foundation proposal. You (and perhaps your team) have spent many hours laying out the details and nuances of your ideas. Now, in developing the summary, you have to change your mindset entirely. Your task in writing an abstract is to provide only the most important, broad-brush elements of your efforts. It must stand on its own and not refer to any of the rest of your proposal. It is, in essence, the "elevator speech" version of your work—only the most important parts of your proposal are mentioned.

The summary/abstract, far from being an easy element to write, is actually one of the most difficult because of the tremendous amount of rethinking you must do to condense the entire proposal into just one paragraph or one page (or whatever the instructions tell you to do). This is the first thing that reviewers see but the last part of your grant application that is written. You can't write the summary until after everything else has been developed.

While instructions will vary depending on the funder, the following is a fairly standard list of elements that should be covered in your summary/abstract. (This particular wording comes from the Office of Community Services but is duplicated in other federal grant RFPs before and since.) "The summary must be clear, accurate, concise, and without reference to other parts of the application. The abstract must include a brief description of the proposed grant project including the needs to be addressed, the proposed services, and the population group(s) to be served" (Office of Community Services, 2013, p. 20).

As you write your abstract, keep your audience in mind. The purpose of any abstract is to introduce and interest your reader in the rest of the proposal. Its purpose has been successfully met if the reviewer finishes reading the summary and wants to find out more. You don't have much space in your summary, and you have many topics to touch on. Still, take time to make the abstract interesting and persuasive—be sure to tell the reader there is a problem and you have a solution.

Kallestinova (n.d.) provides an in-depth set of tips to improve the abstract for any grant, although her focus is on grant proposals for the National Institutes of Health. She cautions authors to avoid wordy abstracts by focusing on the level of vocabulary, level of grammar and level of content. Writers should avoid redundancies, eliminate useless modifiers, and select single words rather than phrases to express themselves. Grammar choices to keep abstracts short include eliminating "overview" phrases, editing "which/that" phrases, and omitting "there is/there are/it is" constructions. Finally, more important content must be prioritized, leaving definitions, details, and data for the body of the proposal.

Titles and abstracts are important to write well, not for the number of points you receive, but for the tone they set. Hurriedly created titles and thoughtlessly written abstracts are easy to spot. The negative impression they give takes considerable effort to overcome. Far better is to plan time to polish these introductory sections of the submission to support, rather than undermine, the remainder of the proposal.

SUBMITTING THE GRANT PROPOSAL

The most important element of submitting your grant proposal is to do it exactly the way the funder wants it submitted. You must read the section of the RFP regarding submitting the grant and follow the directions without variation. If the directions tell you no more than 50 pages and to use 12-point Times New Roman font with double-spaced lines, that is what you must do. Don't even think of using 1.5 line spacing, or 11-point letter size, or Arial font. The funder will toss your proposal into the electronic or old-fashioned trash, and all your hard work will be for naught.

Also, print off a copy of the submission requirements (actually, print off the entire RFP document), read all of the directions and highlight all of the steps. Some of the steps you must do right away even as you're writing the proposal. For example, many federal grants have the following instruction included in the RFP, although this is copied from the grant competition of the Administration on Children, Youth and Families' Grants to Address Trafficking within the Child Welfare Population (US Department of Health and Human Services, 2014, p. 8):

> All applicants must have a DUNS Number (http://fedgov.dnb.com/webform) and an active registration with the Central Contractor Registry (CCR) on the System for Award Management (SAM.gov, www.sam.gov). Obtaining a DUNS Number may take 1 to 2 days.
>
> All applicants are required to maintain an active SAM registration until the application process is complete. If a grant should be made, registration in the CCR at SAM must be active throughout the life of the award. **Finalize a new, or renew an existing, registration at least two weeks before the application deadline.** This action should allow you time to resolve any issues that may arise. Failure to comply with these requirements may result in your inability to submit your application or receive an award.

If you were to wait until the last day before the proposal was due to discover you needed both the DUNS number and the SAM registration, you would not be able to complete the submission.

Because submission requirements for federal government grants vary so much from other entities you may apply to, the only rule to live by is to follow directions. The process of writing and then correctly submitting your proposal is similar to making a loaf of bread from scratch. After you have bought the ingredients, made the effort to knead the dough, spent the time to let the raw loaf rise, and everything else that is required, you don't want to forget to pull the bread out of the oven before it burns. The proposal is not complete until it is submitted correctly to the organization that sets the rules. If you submit the proposal electronically, be sure to save the e-mail receipt showing you were done on time; sometimes the system crashes and you need to have evidence that you had followed correct procedures. If you do not receive an electronic receipt, do not assume that everything went correctly. Follow up immediately with an e-mail or phone call to the project officer in charge of the process to alert him or her to your situation. Similarly, if you are submitting in any other manner, receive proof that you sent in the proposal before the deadline. If your submission process is questioned, the only way to reverse a negative decision is to have such evidence.

HOW THE REVIEW PROCESS WORKS

The most important thing to know about how the review process works is that reviewers compare what you turn in to the explicit guidelines stating what is supposed to be in the proposal. What this means is that, at least for government grants, you have all the information you need to ensure that your proposal is scored well.

A typical review process at the federal level involves the agency receiving the proposals going through a number of steps. (See Box 12.3 for the description provided in one RFP.) First, the agency compiles a list of potential reviewers for that particular RFP. This process occurs before proposals are submitted in order to be ready after the deadline has passed. Reviewers are volunteers who are experts in the area of the RFP. Most reviewers have advanced research degrees, although others are people working in the field. The agency then conducts outreach to see who is available for that round of reviews.

Once selected, the reviewers may receive training in the substance of the RFP they will be judging. One review I participated in required me to take a test over the material in the RFP so that it was clear that I knew what to be looking for when conducting the review. Only after successfully passing the test was I eligible to serve as a reviewer. The review team usually has three reviewers and a chairperson who is responsible for making sure the process is completed in a timely way. The chairperson is responsible for crafting a letter containing committee members' comments about the proposals. Each reviewer reads every proposal and, using the criteria provided, gives a score for every section of the proposals. Each of the proposals may take a couple of hours to read and assess.

At this point, the review team chair facilitates a meeting of the team. Up until recently, review panels met face-to-face for marathon two- and three-day sessions to go over each person's ratings. Currently, with budgets tight, curtailed travel funds, and technology improvements, panels use teleconferencing and video teleconferencing to discuss the proposals from the comfort of their home or office.

BOX 12.3: REVIEW PROCESS INFORMATION FROM COMPETITIVE ABSTINENCE EDUCATION GRANT PROGRAM, HHS-2013-ACF-ACYF-AR-0640

REVIEW AND SELECTION PROCESS

No grant award will be made under this announcement on the basis of an incomplete application. No grant award will be made to an applicant or sub-recipient that does not have a DUNS number (www.dbn.com) and an active registration at SAM (www.sam.gov). See *Section III.3. Other*

INITIAL ACF SCREENING

Each application will be screened to determine whether it meets one of the following disqualification criteria as described in *Section III.3. Application Disqualification Factors*:

- Applications that are designated as late according to *Section IV.3. Submission Dates and Times*,
- Applications that are submitted in paper format without prior approval of an exemption from required electronic submission (*Section IV.2. Request an Exemption from Required Electronic Application Submission*), or
- Applications with requests that exceed the award ceiling stated in *Section II. Award Information*.

For those applications that have been disqualified under the initial ACF screening, notice will be provided by postal mail or by email. See *Section IV.3. Explanation of Due Dates* for information on Grants.gov's and ACF's acknowledgment of received applications.

OBJECTIVE REVIEW AND RESULTS

Applications competing for financial assistance will be reviewed and evaluated by objective review panels using the criteria described in *Section V.1. Criteria* of this announcement. Each panel is composed of experts with knowledge and experience in the area under review. Generally, review panels include three reviewers and one chairperson.

Results of the competitive objective review are taken into consideration by ACF in the selection of projects for funding; however, objective review scores and rankings are not binding. They are one element in the decision-making process.

ACF may elect not to fund applicants with management or financial problems that would indicate an inability to successfully complete the proposed project. Applications may be funded in whole or in part. Successful applicants may be funded at an amount lower than that requested. ACF reserves the right to consider preferences to fund organizations serving emerging, unserved, or underserved populations, including those populations located in pockets of poverty. ACF will also consider the geographic distribution of federal funds in its award decisions.

To be selected to receive a grant, an applicant must demonstrate the following:

- Documented experience in the areas of abstinence and/or adolescent pregnancy prevention education;
- Organizational executive leadership and staffing structure that will support full program implementation within 90 days of grant award;
- A record of demonstrated effectiveness in providing services that are culturally and linguistically relevant to the populations being served.

Source: ACYS (2013, pp. 38–39).

The reviewers all have the same information but often come to different conclusions about how well a proposal meets the RFA's criteria. The role of the chair is to help the reviewers come up with a more uniform view of the strengths and weaknesses of a proposal. Individual reviewers are encouraged and cajoled into modifying their initial scores to reach a common agreement on the worth of each section and the proposal as a whole.

Most federal grants can have a total score of 100, although there are sometimes "bonus points" that can be added for reasons stated in the RFP. If the team gives a score of less than 60, the total score may not even be shared with the grantwriter and applicant organization. Only the comments will be provided. Proposals with higher scores are still in the running for funding.

In a perfect meritocracy, the proposals that had the highest scores would be funded in descending order until the funds ran out. This is not always true, however. As shown in Box 12.3, "objective review scores and rankings are not binding. They are one element in the decision-making process." Other considerations that are important in the final decision include the capacity of the grant-seeking organization to handle the fiscal and managerial demands of having a grant, whether the population to be served is unserved or underserved, and the geographic distribution of the successful applications.

SUMMARY

This chapter covered several topics that relate to parts of the grantwriting process that are important but do not warrant separate chapters. Each topic may make the difference between having your proposal be a winner or not being awarded funding. These topics are sustainability, cultural competency, developing a title and writing the abstract, submission of the grant proposal, and how the review process works. These are truly the "final details" of your grantwriting process. Once the proposal is submitted, you, as the grantwriter cannot influence the process until you are contacted by the funder. It is time to find another RFP to respond to and to start the process once more.

Sustainability asks you to look beyond this particular funding opportunity to plan for a continued program or project. Cultural competency is important so you work with clients in a way that respects them and their cultural and linguistic heritage. A meaningful title and a well-written abstract intrigue your reviewers to want to know more about your ideas and plans. Knowing how to submit your proposal is the cornerstone of being "in the game"—you certainly will not be funded if you've already been rejected for breaking the submission rules. An understanding of the review process can make it less frustrating to endure the waiting for the decision to be made.

If you have followed the guidelines and ideas that have been presented in this book, you will have increased your odds of submitting a winning grant considerably. As noted in this chapter, no guarantees exist in the grantwriting world, even if your proposal is highly rated. That is why it is vital to do everything you can to increase your chances in the Age of Scarcity to be awarded a grant.

Each time you create a new program in response to a foundation or government agency's desire to respond to an existing need you gain additional skill and insight. The reviews that you receive on your proposals may sometimes be scathing (particularly in the beginning of your career) but if you develop the ability to take none of the critique as a personal attack then you have the capacity to learn and improve. While I have had a fair share of success, I have also received the rejections. It is part of the grantwriting profession.

POSTSCRIPT

Any book or training materials on grantwriting are sure to be incomplete and probably outdated just as soon as they are considered finished. Websites change, terminology is modified, and new research and resources emerge requiring fresh understandings. The current funding landscape, which I consider to be in an Age of Scarcity, is not going to make successful grantwriting any easier. In fact, political movements in the United States at the local, state, and federal levels will seek to constrain, if not reduce, spending by governments. Foundations will do what they can to handle increased needs, but they will find their combined resources unable to make up the difference. Individuals will not have enough money to donate so problems cash-strapped nonprofits face can disappear.

Still, this book is written to help those who read it understand what is required to be successful and to give them an edge in harvesting the grant resources that exist. Funding is available, but it is more difficult than ever to win. This book is written to help new grantwriters write their first grant proposals. It is written with experienced fundraisers in mind as well—people who have perhaps not been trained or given a systematic overview of grantwriting from start to finish.

I have used as many examples and relevant documents as I could find. I have given step-by-step instructions that are accurate at the time of their writing. This is both a strength and a potential weakness. Things change with no warning. For example, I posted a video to YouTube explaining how to use www.grants.gov to find federal funding. Just as it reached the top rank for videos on that topic, the federal government changed the look of its website. While the changes were mainly superficial, the video became outdated overnight. Thus, it may be that some of the information in this book is not entirely accurate by the time you read it. Links for RFPs may change (particularly for RFPs that are open at the time of my writing). Other small details may not be correct. But the core messages of the chapters will hold true, and I believe that this book will be a useful resource for you, in all its broader implications, for years to come.

Grantwriting is a skill that people can learn. Not every proposal will be funded, but if you follow the guidelines you've read and practiced while reading this book, you should have a higher than average number of proposals that are selected for funding. The process is logical; the steps are doable.

I wish you the best of luck. I want you to be successful and know that you can be. Remember: Grantwriters write grants. Successful grantwriters write grant proposals that are FUNDED!

PRACTICE WHAT YOU'VE LEARNED

1. Look at an RFP for a grant you are interested in. Does it have requirements for sustainability planning? If so, following the guidelines in the funding announcement, develop a plan for how to maintain the program you are seeking resources for using the suggestions in this chapter. If this is not a requirement for the grant proposal you are currently working on, draft a plan to present to your organization's board following the information in Boxes 12.1 and 12.2. (It is never too early to plan how to maintain a successful program.)
2. Cultural competence should be a goal of every human services agency regardless of funder requirements. Look at the organization you are writing a grant for. Using the requirements listed in this chapter, draft a statement describing how your organization is providing services that are culturally and linguistically appropriate and that meet the Culturally and Linguistically Appropriate Services (CLAS) standards.
3. Here are two actual grant proposal abstracts from different competitions. Read them carefully, noting the elements that are present and what might be missing compared to the information given in this chapter. How might you change them? Develop appropriate and catchy titles for these projects, using no more than 140 characters.

EXAMPLE 1

NOVA SOUTHEASTERN UNIVERSITY, FL

BROWARD COLLEGE, FL

Cooperative Development Grant

ABSTRACT

This Title V Cooperative Arrangement Development application will develop specific programs and services to better meet the needs of Hispanic/Latino and other diverse students who: (1) enter upper division studies as community college transfer students; and (2) are pursuing high-demand science, technology, engineering, and mathematics (STEM) related programs and careers, particularly those related to Computer Science. Funding will support utilizing graduate faculty and model graduate programs to enhance undergraduate transition/support services (including options designed to support student pursuit of graduate- level terminal degrees, if desired), and undergraduate-level STEM degree attainment, specifically targeting Computer Science (CS) and Computer Information Systems (CIS) programming.

The lead institution is Nova Southeastern University (NSU), a highly rated provider of degree opportunities for Hispanic/Latino and other underrepresented students, particularly for those pursuing post-baccalaureate degrees. NSU is nationally ranked in the top 25 institutions for awarding master's degrees (#4), first professional degrees (#2), doctoral degrees (#3), and bachelor's degrees in Biology (#17) to Hispanic/Latinos. The partnership between NSU and feeder Broward College (BC) is natural not only because of close geographic proximity and shared service area, but also because of the existing strong collaborative spirit between the two partners. NSU and BC propose a Cooperative Comprehensive Development Plan to develop: (1) enhanced outreach and academic support services for Hispanic/Latino and other diverse populations that incorporate specific strategies to better serve both native and inter-institutional transfer students; and (2) an inter-segmental, fully-scaffolded degree pathway for Hispanic/Latino and other diverse CS or CIS majors.

A high percentage of Hispanic/Latinos and other underrepresented minorities are enrolled in the NSU CS/CIS programs targeted by this project, making NSU already an institution of access and opportunity; the challenge is to retain and graduate these students in higher numbers. NSU undergraduate faculty and the Office of Undergraduate Student Success will work in close partnership with NSU CS/CIS graduate faculty leaders and with BC to develop and pilot test research-informed best practices that can be sustained and that have been proven to bring about improved student results. Hispanic/Latinos, as well as other underrepresented and low income students, will experience: (1) more rapid progression through the CS/CIS programs via a realigned curriculum; (2) fewer impediments to learning, as students' needs are more proactively anticipated and addressed; (3) sustainable CS/CIS curriculum changes; and (4) an expanded support model that targets native and transfer student needs upon admission and through key degree progression points.

Through collaborative outreach with BC and local industry partners, NSU will create a more accessible and efficient degree pathway that produces well-educated graduates with technical and critical thinking skills necessary for workplace success. NSU will strengthen the rigor of its CS/CIS programs through improved technology, more effective pedagogy including active and project-based learning, and integrated student success strategies such as mentoring. Specific project objectives will focus on increasing enrollment, retention, and graduation rates.

EXAMPLE 2

AUGUSTA TECHNICAL COLLEGE

PR AWARD NUMBER: P382A110011

***State:* Georgia**

The purpose of Augusta Technical College's Predominately Black Institutions (PBI) application is to strengthen programs which improve the educational outcomes of African American male students. Gaps in education achievement signaled to college staff a critical need to focus programs, resources, and data analysis on the African American male student population on their four campuses.

Each activity has been designed to address the goals and objectives of the PBI program, competitive preference priorities and to strategically align with the needs of the college's African American male (AA M) student population:

The Augusta Technical College PBI project will be implemented through three objectives and key strategies:

OBJECTIVE 1: INCREASE THE ACADEMIC SUCCESS OF AA M STUDENTS IN TARGETED DEVELOPMENTAL COURSES BY 4%.

- A writing and math lab entitled the Center for Learning and Academic Support Services—CLASS—will be opened on campus to supplement and target learning in key skills that are often barriers for African American male students who test into developmental level, pre-college course work based on the COMPASS assessment and placement tool.

OBJECTIVE 2: INCREASE THE RETENTION RATE OF AA M STUDENTS BY 4%.

- A Student Success Coordinator will provide targeted and aggressive individual case management and service intervention planning for African American male students. He/She will manage, along with the Learning Support Advisor, the learning support advising, learning support registration and learning support course selection process including identifying and scheduling academic and support services for the AA M student who is not fully engaged in services offered on campus.
- A week-long faculty professional development academy will be offered to best educate and train faculty on developmental education strategies impacting the success of their African American male students. The academy curriculum is designed to increase AA M retention rate by fostering faculty cultural competencies and understanding of the developmental education pedagogy.

OBJECTIVE 3: INCREASE THE COLLEGE'S INFORMATION TECHNOLOGY (IT) AND INSTITUTIONAL RESEARCH (IR) CAPACITY AS DEMONSTRATED BY IMPROVED REPORTING.

- Infrastructure improvements that will support the college in evaluating and analyzing programs serving African American male students.
- Adopt "Achieving the Dream" national strategy to collect and analyze student performance data in order to build a culture of evidence.

4. If your organization does not have a DUNS number, obtain one (with the permission and support of your organization).
5. Determine if your organization has an active registration with the Central Contractor Registry on the System for Award Management, as discussed in the

chapter. If so, write down the information in an easily retrievable location. If not, begin the process to register.

6. Review all other requirements for a typical federal grant, using any of the examples used in this book. Would your organization be able to upload the documents successfully, per the description? Find a foundation that might want to fund your proposal. What does it demand in terms of submitting a proposal? Would you have any problems with fulfilling those requirements?
7. Research one or more federal agencies that use volunteer grant reviewers. Place yourself on the list to become a reviewer and serve if asked. There is nothing like going through the grants process from the reviewer's side to help you understand the process in a new light.

NOTE: Grantwriters always want to see examples of successful federal grants in order to determine whether their proposals are in the same ballpark. Most agencies are reluctant to provide their proposals (successful or not) to other agencies or the public at large. The National Association for the Mentally Ill (NAMI) provides access to a successful federal grant from the Substance Abuse and Mental Health Services Administration (SAMHSA) which you may view at: http://www.nami.org/Content/ContentGroups/Policy/CIT/NAMI_Maine_SAMHSA_GRANTS.pdf

REFERENCES

Administration for Children and Families (2012). CCF/SCF Tools Creating Your Sustainability Plan. Retrieved from www.acf.hhs.gov/programs/ocs/resource/creating-your-sustainability-plan

Hoefer, R. (2012). From website visitor to internet contributor: Three internet fundraising techniques for non-profits. *Social Work*, 57(4): 361–365. doi: 10.1093/sw/sws002

Kallestinova, E. (n.d.). How to Write a Winning Abstract for Grant Applications. Retrieved from http://www.yale.edu/grants/funding_info/pdf/Abstracts.pdf

Office of Community Services (2013). Community Economic Development Projects. HHS-2013-ACF-OCS-EE-0583. Retrieved from http://www.acf.hhs.gov/grants/open/foa/files/HHS-2013-ACF-OCS-EE-0583_1.pdf

Office of Minority Health (2013). The National CLAS Standards. Retrieved from http://minorityhealth.hhs.gov/templates/browse.aspx?lvl=2&lvlID=15

US Department of Health and Human Services, Administration for Children and Families, Administration on Children, Youth and Families (2013). Competitive Abstinence Education Grant Program HHS-2013-ACF-ACYF-AR-0640. Retrieved from https://ami.grantsolutions.gov/files/HHS-2013-ACF-ACYF-AR-0640_0.htm

US Department of Health and Human Services, Administration for Children, Youth and Families (2012). Teen Drug Court. RFA TI-12-004. Retrieved from http://apply07.grants.gov/apply/opportunities/instructions/oppTI-12-004-cfda93.243-instructions.pdf

US Department of Health and Human Services, Administration for Children and Families, Administration on Children, Youth and Families (2014). Grants to Address Trafficking Within the Child Welfare Population. HHS-2014-ACF-ACYF-CA-0831. Retrieved from http://www.acf.hhs.gov/grants/open/foa/view/HHS-2014-ACF-ACYF-CA-0831

US Department of Health and Human Services (US DHHS), Substance Abuse and Mental Health Services Administration (2012). Teen court program. Request for Applications (RFA). No. TI-12-004. http://apply07.grants.gov/apply/opportunities/instructions/oppTI-12-004-cfda93.243-instructions.pdf

INDEX

CPSIA information can be obtained
at www.ICGtesting.com
Printed in the USA
BVHW011554020921
615260BV00013B/16